A Programmed Introduction to Number Systems

A Programmed Introduction to Number Systems

SECOND EDITION

Irving Drooyan
Walter Hadel
Los Angeles Pierce College

John Wiley & Sons, Inc.
New York   London   Sydney   Toronto

Copyright © 1964, 1973, by John Wiley & Sons, Inc.

All rights reserved. Published simultaneously in Canada.

No part of this book may be reproduced by any means, nor transmitted, nor translated into a machine language without the written permission of the publisher.

ISBN 0-471-22266-6

Printed in the United States of America

10 9 8 7 6 5 4 3 2 1

Preface

This revision of *A Programmed Introduction to Number Systems* reflects eight years of students' experience with the first edition of the program. Textual material has been rewritten to improve the presentation and the exercise sets have been expanded to supplement the program for additional practice on skills. In addition, a new unit has been added which includes material on irrational numbers and the extension of the set of rational numbers to form the real number system.

Several improvements have been made in the format to facilitate the use of the book. A list of symbols, introduced in the program, is printed on the back of the perforated response shield. Panel IV has also been perforated so that it can be easily removed from the text. These two detachable cards can be used by students as convenient references for their further studies.

As the title indicates, this programmed text provides an introduction to number systems, a topic which is fundamental to many subjects in the mathematics curriculum. This topic provides a background in the structure of both arithmetic and algebra by emphasizing the recurrent patterns in both subjects. The table of contents and the detailed objectives for each part of the program show the specific topics discussed.

This text can be used to good advantage in different ways:

1. For students in algebra classes who have not had a previous opportunity to become acquainted with the ideas pertaining to number systems.
2. As a supplementary text in beginning algebra classes in which traditional texts are used.
3. For in-service training of teachers of grades K through 8 in institutes, workshops, or on an individual basis.
4. As a supplement to basic texts in pre-service training for teachers and texts used in mathematics courses oriented toward general education (see bibliography on page 240).

The only prerequisites required to succeed in the program are an interest in the subject and the necessary time to complete the program.

The average time required to complete each part of the program, based on the use of the material by a variety of students, is shown in the following table.

Unit I	2 hours	30 minutes
Unit II	5 hours	45 minutes
Unit III	2 hours	50 minutes
Unit IV	3 hours	50 minutes
Unit V	2 hours	5 minutes
Total	17 hours	

Instructions on how to use the programmed material in this text most effectively for both individual study and classroom use follow the table of contents.

We express our thanks to the many students who worked through the preliminary forms of this program and to their instructors who offered numerous suggestions, many of which were incorporated into the text.

Woodland Hills, California

Irving Drooyan
Walter Hadel

How to Use This Material

This program presents information in two ways. First, you will find paragraphs of written exposition, each entitled "Remark," which vary in length from a few lines to a page or more. Second, you will find small units of information, or short questions (called "frames"), each of which requires you to make an active response in the form of a word, a phrase, or a symbol. Both the remarks and the frames are important in achieving the objectives of the program.

To progress through the program, begin on page 1, and:

1. Place a shield over the page so that only the beginning remark and the first frame are exposed.
2. Read the introductory remark.
3. Read the first frame carefully, noting as you do the place where you are asked to respond.
4. After deciding on the proper response, write it on an auxiliary response sheet.
5. Slide the response shield down until it is level with the top of the next frame. This will uncover the correct response to the preceding frame and any additional remarks accompanying it.
6. Verify that you have made the correct response. If you have, go on to Step 7. If you have not, reread the frame with the correct response in mind, and then write the correct response beside (or above or below) your earlier incorrect response. Then go on to Step 7.
7. Read and respond to the next frame.
8. Continue to repeat Steps 5, 6, and 7 until you have completed the program. When you encounter a "Remark," slide the shield down just far enough to expose the complete remark.
9. When turning pages, try to avoid looking at the material on the new page until you have covered any responses with the shield.

Essentially, these steps form your part of the task; the program must assume the remainder of the responsibility for your learning. However, there are a few additional things you can do to improve the effectiveness of the material:

vii

1. Read the remarks and the frames very carefully. If you are reading a frame, consider your response before writing it. Does it fit the wording of the frame? Does it make sense? *Do not try to go too fast.*

2. Mathematics is a language written in symbols. In order to understand written mathematics, you have to make a conscious effort to read the symbolism as though written in words. This is extremely important in self-instructional material. If you are not very careful in this regard, the symbolism can quickly become meaningless to you. Every symbol or group of symbols used in this program has a meaning which can be interpreted in words, and you should know this meaning.

3. Do not try to do too much at once. If you stay with your work too long at one time, you tend to become impatient, and the number of errors you make will probably rise. After taking a break, pick up your work a few frames earlier than where you stopped.

4. If you want additional reenforcement with respect to certain parts of the program, you can complete the Supplemental Exercises that are provided and noted at intervals in the program.

Contents

How to Use This Material — vii

UNIT I	Preliminary Concepts	2
UNIT II	The Set of Whole Numbers	30
UNIT III	The Set of Integers	106
UNIT IV	The Set of Rational Numbers	146
UNIT V	The Set of Real Numbers	212

APPENDICES

Bibliography	240
Table of Square Roots	241
Supplemental Exercise Sets	242
Answers to Self-evaluation Tests	278
Index	281

Response Shield

A Programmed Introduction to Number Systems
SECOND EDITION

Irving Drooyan and Walter Hadel

Use this shield to cover the response and
all frames below the one that you are working.

$\{$ natural numbers $\}$

▼

$\{$ whole numbers $\}$

▼

$\{$ integers $\}$

▼

$\{$ rational numbers $\}$
union
$\{$ irrational numbers $\}$

▼

$\{$ real numbers $\}$

The page where the symbol is first used is shown in parentheses.

$A, B, C, \ldots$	names of sets (3)
$\{a, b\}$	the set whose elements (or members) are a and b (4)
$=$	is equal to (4)
$\neq$	is not equal to (5)
$\in$	is an element of or is a member of (5)
$\notin$	is not an element of or is not a member of (5)
$\emptyset$	the null set or the empty set (6)
$\subset$	is a subset of (7)
$\cup$	the union of (12)
$\cap$	the intersection of (13)
U	the universal set (14)
A'	complement of A (15)
(a, b)	an ordered pair (23)
$A \times B$	Cartesian product of A and B (23)
N	the set of natural numbers (34)
W	the set of whole numbers (34)
$<$	is less than (35)
$>$	is greater than (35)
$a, b, c, \ldots$	variables a, b, c, etc. (38)
$\{a \mid \ldots\}$	the set of all a such that $\ldots$ (39)
$a + b$	the sum of a and b (51)
$a \times b$ or $a \cdot b$	the product of a and b (62)
$a - b$	the difference of b subtracted from a (85)
$a \div b$ or $\frac{a}{b}$	the quotient of a divided by b (89)
$-1, -2, -3, \ldots$	negative integers (107)
$-a$	the additive inverse of a or negative of a (108)
J	the set of integers (111)
$\lvert a \rvert$	the absolute value of a (120)
$\frac{1}{b}$	the multiplicative inverse of b or reciprocal of b (147)
Q	the set of rational numbers (151)
$\geq$	is greater than or equal to (214)
$\sqrt{b}, b \geq 0$	the non-negative square root of b (214)
H	the set of irrational numbers (216)
R	the set of real numbers (216)

UNIT I
Preliminary Concepts

OBJECTIVES

Upon completion of this part of the program, the reader will be able to:

1. Use the correct vocabulary and/or the correct notation to express the concepts of sets, elements of sets, equal sets, null set, disjoint sets, the intersection and union of sets, the universe, the complement of a set, an ordered pair, and the Cartesian product of two sets.

2. Construct Venn diagrams to illustrate a subset, the intersection of two sets, the union of two sets, disjoint sets, and the complement of a set.

Remark. As the title of this program implies, this is a brief introduction to the nature and properties of number systems. A number system involves things other than numbers; it involves some operations on the numbers and some rules governing the operations. To begin with, we want to consider the notion of a set—a topic that pervades wide areas of mathematics.

1. Any collection of things considered together is called a **set**. A senior class and a baseball team are examples of sets. The letters in the English alphabet may be said to form a _____.

   ---
   set

2. Any one of the things contained in a set is called a **member** of the set. For example, you are a member of the _____ of people in the United States.

   ---
   set

3. Another word for "member" is "**element**." q is a _____ or an _____ of the set of letters in the English alphabet.

   ---
   member; element

4. The boys in a class are members or _____ of the _____ of all students in the class.

   ---
   elements; set

5. Capital letters such as A, B, and C are used to denote sets. For example, C might be used to denote the set of vowels in the English alphabet. In this case, a, e, i, o, u, and y would be _____ of the set C.

   ---
   members (*Or elements.*)

6. Any capital letter, for example G, might be used to denote the set of students who scored 100% on a mathematics test. Thus, if Mary, John, and Bill scored 100% on this test, then Mary, John, and Bill would be _____ or _____ of the set G.

   ---
   members; elements

7. Braces, { }, are sometimes used to enclose the members of sets. {a, e, i, o, u, y} is read "the set whose members are a, e, i, o, u, and y." {Mary, John, Bill} is read "the _____ _____ _____ _____ Mary, John, and Bill."

---
set whose members are

8. The order in which the elements of a set are listed is *usually* of no importance. Thus, {a, b} could be written {_____}.

{b, a}

Remark. In the previous frame, the braces denoting a set were included in the response because the line in the frame used to indicate a response extended beyond the braces.

9. The members of a set may be described *by using a rule* or *by listing the members*. If Mary, John, and Bill are all of the students in a class who scored 100% on a certain test, then {students in the class who scored 100% on the test} and {_____ , _____ , _____} both represent the same set.

{Mary, John, Bill} *(Any other order can be used to list these names.)*

10. If the members of {vowels in the English alphabet} were to be listed, the representation would appear as {_____}.

{a, e, i, o, u, y}

11. Two sets, A and B, are **equal** if all members of A are members of B and all members of B are members of A. Thus {vowels in the English alphabet} and {a, e, i, o, u, y} are equal _____.

---
sets

12. {Jane, Sue, Janet} and {Sue, Janet, Jane} are _____ sets.

---
equal

13. The symbol, =, represents the word "equals" or the words "is equal to." The statement $A = B$ means that A and B are _____ sets.

---
equal

Preliminary Concepts 5

14. If $A = \{$Mary, John, Bill$\}$, and $B = \{$Bill, Mary, John$\}$, then A ____ B.

$A = B$ (*Members of equal sets do not have to be listed in the same order.*)

15. The symbol, $\neq$, represents "is not equal to." If $A = \{$Mary, John, Bill$\}$, $B = \{$Bill, Mary, John$\}$, and $C = \{$Mary, John, Jim$\}$, then which of the following are true statements?

$$A = B, \quad A \neq B, \quad A = C, \quad A \neq C, \quad B = C, \quad B \neq C$$

$A = B; \; A \neq C; \; B \neq C$

16. If $E = \{\bigcirc, *, \square\}$, $F = \{\triangle, *, \bigcirc\}$, and $G = \{*, \square, \bigcirc\}$, then F ____ G and E ____ G.

$F \neq G; \; E = G$

17. The symbol epsilon, $\in$, represents the phrase "is a member of" or "is an element of." If $A = \{$Mary, John, Bill$\}$, then the symbols "Mary $\in A$" are read "Mary is __ _____ of A."

---

a member (*Or an element.*)

18. If $B = \{a, e, i, o, u, y\}$, the fact that y is a member of the set can be expressed symbolically by writing y ____ B.

$y \in B$

19. The fact that i is a member of B in the preceding frame can be expressed concisely by writing _____.

$i \in B$

20. If $W = \{$days of the week$\}$, express the fact that Monday is a member of the set.

Monday $\in W$ (*Or* Monday $\in \{$days of the week$\}$.)

21. The symbol $\notin$ represents "is not an element (or a member) of." For example, we might write $r \notin \{$vowels in the English alphabet$\}$. However, r ____ $\{$letters in the English alphabet$\}$.

$\in$

22. January _____ {days of the week}.

∉

23. Bill ∉ {high school students} is read "Bill ___ ___ ___ _____ ___ _____
    _____ ___ high school students."

------

is not a member (or an element) of the set of

24. If $S = \{a, e, i, o, u, y\}$, then e _____ S and x _____ S.

$e \in S$; $x \notin S$

25. If $Q = \{$Susan B. Anthony, Mata Hari, Seabiscuit$\}$, then Mata Hari _____ Q and John Quincy Adams _____ Q.

$\in$; $\notin$

Remark. The whole point to the preceding frame is that the members of a set do not need to have characteristics in common other than membership in the set. Of course, such sets are generally not of much interest to anyone, and the sets with which we are concerned in mathematics are usually not of this type. The members of the sets dealt with in mathematics are almost always alike in some significant way, and in a well-formed set, it should always be possible to determine whether something is or is not in the set. The set symbolism introduced thus far consists of capital letters and { } to denote sets, $\in$ and $\notin$ to denote membership and nonmembership, and = and ≠ to denote "is equal to" and "is not equal to."

The sets that we have encountered have been of the form $\{a, b, c\}$, $\{$Mary$\}$, and $\{\bigcirc, *, \square, \triangle\}$. Let us look at some additional symbols used in working with sets.

26. It is possible to conceive of a set that contains no members. For example, {women who have been elected President of the United States} contains _____ members.

------

no

27. The set that contains no members is called the **empty set** or the **null set**, and is denoted by the symbol ∅ (read "the empty set" or "the null set"). Thus, {cows that speak perfect English} can be denoted by _____ .

∅

Preliminary Concepts

Remark. Braces are not used with the symbol ∅. The symbol { }, however, is sometimes used instead of the symbol ∅ to denote the empty set. We shall use ∅ in this program.

28. If no students in a class scored below 50% on a test, {students who scored below 50%} is the empty set or _____ set and is represented by the symbol _____.

    null; ∅

29. Each set has one or more **subsets**. A is a subset of B if each member of A is also a member of B. Thus, $\{e, i\}$ is a subset of $\{a, e, i, o, u, y\}$ because e and i are _____ of both sets.

    members *(Or elements.)*

30. {Students in the Freshman class at Taft High School} is a _____ of {students in Taft High School}.

    subset

31. Each set is a subset of itself. Thus $\{a, e, i, o, u, y\}$ is a subset of $\{a, e, i, o, u, y\}$ and {students in high school} is a _____ of {students in high school}.

    subset

32. The empty set is considered a subset of every set. Hence it is a subset of itself. Thus, _____ is a subset of ∅.

    ∅

33. The symbol ⊂ is used to represent "is a subset of." $A \subset B$ asserts that A is a subset of B. $G \subset H$ asserts that G is a _____ of H.

    subset

34. $P \subset Q$ asserts that ___ is a subset of ___.

    P; Q

UNIT I

35. If $A = \{*, \triangle, \bigcirc\}$, $B = \{*, \triangle\}$, and $C = \{*\}$, then which one of the following is a true statement?

$$A \subset B \;/\; B \subset C \;/\; C \subset A$$

$C \subset A$

36. If $A = \{a, b, c\}$, $B = \{a, b\}$, and $C = \{b, c\}$, then which one of the following is *not* a true statement?

$$B \subset A \;/\; B \subset C \;/\; C \subset A$$

$B \subset C$

37. Express in symbols the facts that G is a subset of H, and K is a subset of G.

$G \subset H;\; K \subset G$

38. Since each set is a subset of itself, express in symbols the fact that $\{*, \square\}$ is a subset of $\{*, \square\}$.

$\{*, \square\} \subset \{*, \square\}$

Remark. It is sometimes easy to confuse the meaning of the symbols $\subset$ and $\in$. When we write $a \in A$, we mean that a is an *element* of the *set* A. However, if we write $\{a\} \subset A$, we mean something else, namely, that the *set* whose only member is a is a *subset* of A. The symbol $\in$ applies to *individual members* of sets, while $\subset$ applies to *subsets* of sets.

39. Let $U = \{a, b\}$, then $a \in U$ and $b \underline{\quad\quad} U$.

$b \in U$

40. Select the two true statements. If $U = \{a, b\}$, then

$$\{b\} \subset U \;/\; \{b\} \in U \;/\; b \subset U \;/\; b \in U$$

$\{b\} \subset U;\; b \in U$

41. a, b, and c are (members/subsets) of $\{a, b, c\}$.

---
members

Preliminary Concepts

42. Select the two true statements. If $B = \{a, c, e\}$, then
$$\{a\} \subset B \;/\; \emptyset \subset B \;/\; \{c\} \in B$$

$\{a\} \subset B$; $\emptyset \subset B$

43. The set that is a subset of every set is denoted by _____ .

$\emptyset$

44. $\{\triangle\}$ and $\{\bigcirc\}$ are (elements/subsets) of $\{\triangle, \bigcirc\}$.

---
subsets

45. The four subsets of $\{a, b\}$ are $\emptyset$, $\{a\}$, $\{\;\;\}$, and $\{\;\;,\;\;\}$.

$\{b\}$; $\{a, b\}$

46. The three subsets of $\{a, b, c\}$ that contain only one member are $\{\;\;\}$, $\{\;\;\}$, and $\{\;\;\}$.

$\{a\}$; $\{b\}$; $\{c\}$

47. Represent the subset of $\{a, b, c\}$ that contains no members.

$\emptyset$

48. List the eight subsets of $\{a, b, c\}$.

$\emptyset$, $\{a\}$, $\{b\}$, $\{c\}$, $\{a, b\}$, $\{a, c\}$, $\{b, c\}$, $\{a, b, c\}$

49. The elements of two sets, A and B, can be placed in **one-to-one correspondence** if each element of A corresponds to one and only one element of B and each element of B corresponds to one and only one element of A. The symbol $\updownarrow$ indicates those elements in the two sets that have been placed in one-to-one correspondence to each other.

$$\text{If } A = \{\; e,\quad i,\quad o,\quad u,\quad v,\;\}$$
$$\quad\quad\quad\updownarrow\quad\updownarrow\quad\updownarrow\quad\updownarrow\quad\updownarrow$$
$$\text{and } B = \{\text{book, chair, hand, dog, table}\},$$

▼

▼(This symbol indicates that the frame or remark continues on the next page.)

10   UNIT I

the members of the sets are in _____ _____ _____ correspondence. The element u in A corresponds to the element _____ in B.

---

one-to-one; dog

Remark. The one-to-one notion here amounts to a pairing of the elements of the two sets, where no element in either set is paired with more than one element in the other.

50. Let $B = \{$bed, chair, table$\}$
    $\updownarrow$ $\updownarrow$ $\updownarrow$
    and $C = \{\ a,\ \ \ b,\ \ \ c\ \}$.

    The members of sets B and C are in one-to-one _____.

---

correspondence

51. Let $R = \{$catcher, pitcher, umpire, coach$\}$
    $\updownarrow$ $\updownarrow$ $\updownarrow$
    and $S = \{\ \ 1,\ \ \ 2,\ \ \ 3,\ \ \ \}$.

    The members of sets R and S are *not* in _____ _____ _____ _____.

---

one-to-one correspondence

52. Two sets whose members can be placed in one-to-one correspondence are **equivalent.** $\{$dog, cat, horse, cow$\}$ and $\{$April, May, June$\}$ (are/are not) equivalent.

---

are not

53. Consider $G = \{$Richard, John, Bill$\}$ and $S = \{$Karen, Susan, Doris$\}$. Because the members of sets G and S can be placed in one-to-one correspondence, the sets are _____.

---

equivalent (*They are not equal sets because the elements are not the same.*)

54. $\{\bigcirc\}$ and $\{*\}$ are equivalent because their members can be placed in _____ _____ _____ .

---

one-to-one correspondence

Preliminary Concepts

55. {Jack, Terry} and {Robert, Russell} are (equal/equivalent) sets.

---

equivalent

Remark. Note that equivalence does not mean equality. The two concepts are different. If two sets are equal (equality), then they contain exactly the same things as members. Equivalence, however, implies only that the members of one set can be paired one-to-one with the members of another, and these members need not even be the same kinds of things.

Sometimes two sets do not contain common members; other times they share some, but not all, of their members. Let us examine these notions.

56. Two sets, A and B, are **disjoint** if they contain no members in common. Thus, {horses} and {people} (are/are not) disjoint.

---

are

57. If the letters $a, b, c, d,$ and e represent distinct (different) things, then the sets $\{a, b, c\}$ and $\{d, e\}$ are _____ .

---

disjoint (*They are also unequal.*)

58. $\{a, b, c\}$ and $\{c, d, e\}$ are not disjoint because they each contain ____ as a member.

---

c

59. $\{\Box, \bigcirc, *, \triangle\}$ and $\{*, \Box\}$ (are/are not) disjoint.

---

are not

60. If $A = \{$boys in the second grade$\}$ and $B = \{$girls in the second grade$\}$, then A and B are _____ .

---

disjoint

61. Assume each letter denotes a distinct thing; list the disjoint sets.

$$\{e, f, g\}, \{g, b, i\}, \{c, g, i\}, \{b, c, d\}$$

$\{e, f, g\}$ and $\{b, c, d\}$

UNIT I

Remark. See Exercise Ia, page 242, for additional practice on the topics that have been introduced to this point.

We shall now consider several operations on sets that enable us to construct new sets in terms of given sets.

62. The set consisting of all those elements that are members of either one or the other of two sets A and B is called the **union** of A and B. If $A = \{\text{Mary, Jane}\}$ and $B = \{\text{Sally, Donna, Sue}\}$, then $\{\text{Mary, Jane, Sally, _____, _____}\}$ is the union of A and B.

Donna, Sue

63. If $A = \{a, b, c\}$ and $B = \{b, c, d\}$, then $\{a, b, c, d\}$ is the _____ of A and B.

----

union

Remark. Notice that, in listing the elements of a set, the same symbol is not used more than once. This is because, conceptually, the referent occurs but once in the set, and repeated use of the same symbol is redundant. Thus, the elements of $\{a, a, b, c, d\}$ are $a, b, c,$ and d, and no point is served in listing a twice.

64. The union of two sets A and B is denoted by the symbols $A \cup B$, which is read "the union of A and B." Thus $\{\triangle, \square, \bigcirc\} \cup \{*, \bigcirc\} = \{\text{_____}\}$.

$\{\triangle, \square, *, \bigcirc\}$

65. If $A = \{a, b, c\}$ and $B = \{d, e\}$, then $A \cup B = \{\text{_____}\}$.

$\{a, b, c, d, e\}$

66. For any set A, $A \cup A = $ _____.

A

67. For any set A, $A \cup \emptyset = $ _____.

A

Preliminary Concepts 13

68. Consider a set consisting of all elements that are members of both *A* and *B*. Such a set is called the **intersection** of the given sets. If $A = \{a, b, c\}$ and $B = \{b, c, d\}$, then $\{b, c\}$ is the intersection of *A* and *B* because ____ and ____ are elements of both *A* and *B*.

---
b; c

69. If $A = \{\bigcirc, \triangle, \square\}$ and $B = \{\square, *, \bigcirc\}$, then $\{\bigcirc, \square\}$ is the _____ of *A* and *B* because $\bigcirc$ and $\square$ are common members of *A* and *B*.

---
intersection

70. The intersection of $\{$Mary, Bill, John$\}$ and $\{$Jane, Richard, Bill$\}$ is _____.

$\{$Bill$\}$ (*Note that the intersection of the given sets is another set $\{Bill\}$ and not simply the element "Bill."*)

71. The intersection of $\{$Mary, Bill$\}$ and $\{$Jane, Richard, John$\}$ is the _____ set.

---
null (*Or empty.*)

72. The intersection of two sets, *A* and *B*, is denoted by the symbols $A \cap B$, which is read "the intersection of *A* and *B*." Therefore, $\{$Bill, John$\} \cap \{$Jane, John$\}$ = $\{$_____$\}$.

$\{$John$\}$

73. If $A = \{a, e, i, y\}$ and $B = \{e, o, u\}$, then $A \cap B$ is ____.

$\{e\}$ (*Again, the intersection of two sets is a set.*)

74. $\{\triangle, *, \square\} \cap \{*, \triangle, \square\}$ = _____.

$\{\triangle, *, \square\}$

75. For any set *A*, $A \cap A$ = ____.

A

14 UNIT I

76. $\emptyset \cap \emptyset =$ ____ ; $A \cap \emptyset =$ ____.

$\emptyset$; $\emptyset$

77. {boys in the twelfth grade} $\cap$ {girls in the twelfth grade} = ____.

$\emptyset$ (*The sets are disjoint.*)

78. For all disjoint sets A and B, $A \cap B =$ ____.

$\emptyset$

79. If $R = \{\triangle, \square, \bigcirc\}$ and $S = \{*, \bigcirc\}$, then $\{\bigcirc\}$ represents the intersection of R and S and $\{\triangle, \square, *, \bigcirc\}$ represents the ____ of R and S.

union

80. $A \cap B$ is read "the ____ of A and B," and $A \cup B$ is read "the ____ of A and B."

intersection; union

81. If $A = \{$Bill, Robert, John, Tom$\}$ and $B = \{$Robert, Tom, Jim$\}$, then $A \cap B = \{$ ____ $\}$ and $A \cup B = \{$ ____ $\}$.

{Robert, Tom}; {Bill, Robert, John, Tom, Jim}

82. If $A = \{$John$\}$ and $B = \{$Bill, Henry$\}$, then $A \cap B =$ ____ and $A \cup B =$ ____.

$\emptyset$; {John, Bill, Henry}

83. If $A = \{*, \triangle, \bigcirc\}$ and $B = \{\square, *, \bigcirc\}$, then $A \cap B =$ ____ and $A \cup B =$ ____.

$\{*, \bigcirc\}$; $\{\square, *, \triangle, \bigcirc\}$

84. Suppose we wish to consider the Freshman students in Taft High School. Then, the set of all these students is called the **universe of discourse** or simply the

▼

Preliminary Concepts 15

universe and is generally designated by the capital letter U. In this example
$U = \{$ _____ in Taft High School $\}$.

{Freshman students in Taft High School}

Remark. Note that the symbol U denotes the universe, while $\cup$ denotes the union of two sets. The symbols, like the ideas, are different.

85. The statement $U = \{$Sophomores$\}$ asserts that the set of all sophomores is the _____.

---

universe

86. If $U = \{$Sophomores$\}$ and $G = \{$Sophomore girls$\}$, then $G \subset U$. If $B = \{$sophomore boys$\}$, then B ____ U. The set $B \cup G$ is the _____.

$B \subset U$; universe

87. If G is a subset of U, the set of all members of U that are not in G is called the **complement** of G and designated by $\sim G$, $\overline{G}$, or G' (we will use G'). Thus, if $U = \{$sophomores$\}$ and $G = \{$sophomore girls$\}$, then $G' = \{$ _____ $\}$.

{sophomore boys}

88. If $U = \{$students$\}$ and $R = \{$students who wear glasses$\}$, then $R' = \{$students who do not wear glasses$\}$, and R' is called the _____ of R.

---

complement

Remark. In considering sets, we like to have in mind some general things that will be members of our sets. For instance, when we want to consider a set of students, we might have in mind, say, all of the students in a given school, or in a given school district, or we might even want to talk about all students everywhere. The point is that the members of a set can be drawn from a number of different populations, called universes, and it is necessary to specify the universe from which you are drawing the members of the sets unless it is obvious from the context. If a universe U contains a set A, then everything in the universe is either in A or in its complement A', but not in both. *A set and its complement are always disjoint.* We shall present a few more frames on this idea. When reading the frames, it is advisable to read the symbolism as though

▼

16   UNIT I

written out in words. Thus, "If $a \in A$ and $A \subset U$, then $a \in U$" should be read "If a is a member of A and A is a subset of U, then a is a member of U." This can be thought of as something akin to "If a is in the set A, and if everything in the set A is in the set U, then a has to be in U." In short, make the symbols mean something to you as you read them.

89. If a discussion is limited to vowels in the English alphabet, then $\{a, e, i, o, u, y\}$ is the _____ of discourse.

----

universe

90. If $U = \{a, e, i, o, u, y\}$ and $A = \{a, i\}$, then $A' = \{$ _____ $\}$.

$\{e, o, u, y\}$

91. If $U = \{\square, *, \bigcirc, \triangle\}$ and $A = \{\square\}$, then $A' = \{*, \bigcirc, \triangle\}$. In this example, $\{\square, *, \bigcirc, \triangle\}$ is the _____ , $A \subset U$, and A' is the _____ of A.

----

universe; complement

92. If $A \subset U$, and if $a \in A$, then a ($\in$/$\notin$) A'.

$\notin$ (a cannot be in both A and its complement A'.)

93. If $A \subset U$, $a \notin A'$, and $a \in U$, then which of the following statements is correct?

$$a \in A \quad / \quad a \notin A$$

$a \in A$ (a has to be in one or the other but not both.)

Remark. You may more readily understand the foregoing notions of set, subset, intersection, union, universe, and complement through visual representations called **Venn diagrams**, named after their originator. The universe with which we are dealing is usually represented by a rectangle (though any closed plane geometric shape may be used), thus:

U

Preliminary Concepts 17

Subsets are represented by other closed plane geometric figures within the rectangle. Circles are convenient to use for this purpose. Let us see how we can visualize those ideas already discussed.

94. If $U = \{\text{four-legged animals}\}$ and $A = \{\text{lions}\}$, a Venn diagram illustrating that $A \subset U$ would appear as

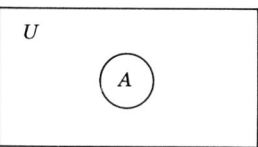

If $U = \{\text{horses}\}$ and $B = \{\text{white horses}\}$, a Venn diagram illustrating that $B \subset U$ could be shown as

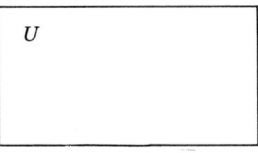

(Complete diagram)

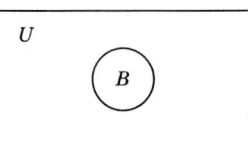

Remark. Any closed plane geometric figure can be used to represent the set B in the preceding frame. The size of the figure does not imply the number of elements that are in a set.

95. If $U = \{\text{students}\}$, $A = \{\text{Freshman students}\}$, and $B = \{\text{Freshman girls}\}$, a Venn diagram illustrating possible relationships would appear as follows:

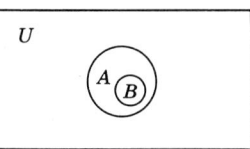

Which one of the following statements is *not* true?

$A \subset U \quad / \quad B \subset U \quad / \quad A \subset B \quad / \quad B \subset A$

$A \subset B$

96. The Venn diagram

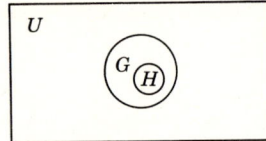

illustrates that *H* is a _____ of *G*.

---
subset

97. A Venn diagram illustrating two disjoint sets in a universe *U* is shown by

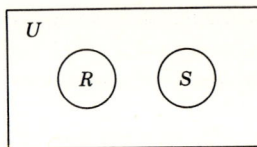

If *U* = {vegetables}, *M* = {carrots}, and *N* = {beets}, a Venn diagram would be:

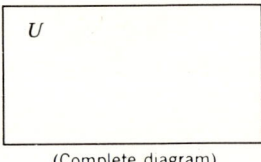

(Complete diagram)

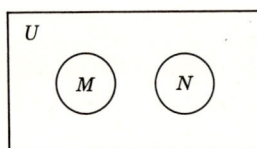

98. If two sets, *A* and *B*, have some members in common, a Venn diagram would appear:

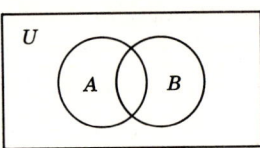

The intersection of *A* and *B* can be shown by shading those regions of *A* and *B* that overlap. Thus, the intersection of *A* and *B* would appear:

▼

Preliminary Concepts

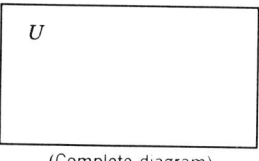
(Complete diagram)

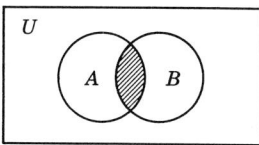 $(A \cap B)$

99. The union of sets A and B in Frame 98 can be shown by shading the entire space occupied by both A and B, and would appear as

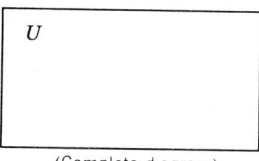
(Complete diagram)

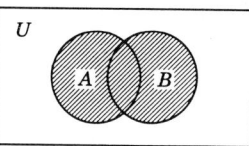 $(A \cup B)$

100. If $U = \{\text{girls}\}$, $B = \{\text{blonde girls}\}$, and $D = \{\text{Mary (brunette), Sue (blonde), Jane (redhead), Kay (brunette)}\}$ a Venn diagram for $B \cap D$ would be:

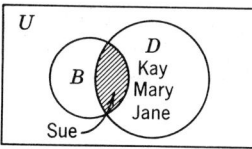

Make a Venn diagram showing $B \cup D$.

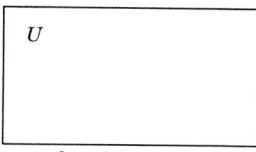
(Complete diagram)

▼

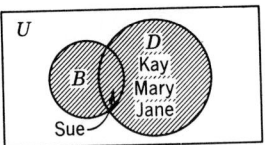

101. If disjoint sets A and B are represented in a Venn diagram as

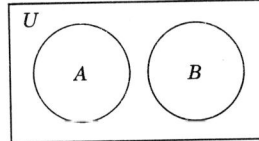

$A \cup B$ would be represented by

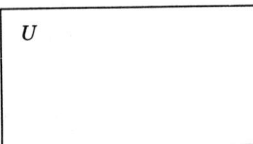

(Complete diagram)

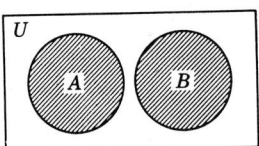

102. A Venn diagram for A', the complement of A, would be represented by

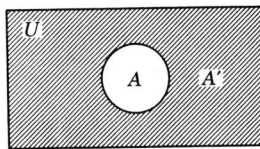

Shade the region representing $(C \cup D)'$, the complement of $C \cup D$.

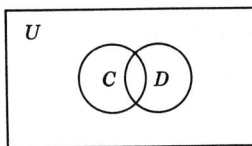

▼

Preliminary Concepts

21

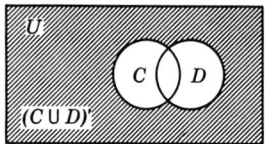

103. Shade the region representing $C \cap D$.

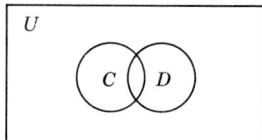

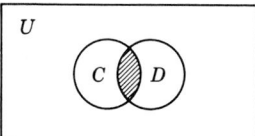

104. The shaded region of the Venn diagram

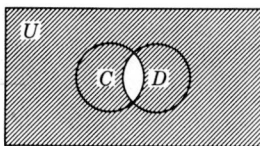

depicts the _____ of $C \cap D$.

---
complement

105. If $U = \{a, b, c, d, e, f, g, h, i\}$, $M = \{c, d, e\}$, and $N = \{e, f, g, h\}$, show a Venn diagram for $M \cup N$. List the elements and shade the appropriate region.

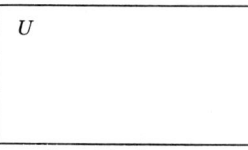

(Complete diagram)

▼

22   UNIT I

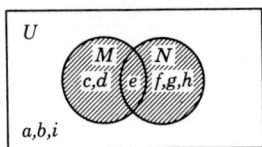

$(M \cup N = \{c, d, e, f, g, h\})$

106. Show a Venn diagram depicting $M \cap N$ for the sets M and N as given in Frame 105.

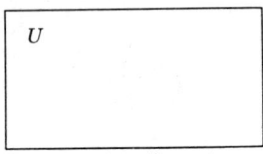

(Complete diagram)

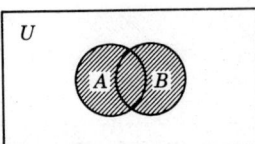

$(M \cap N = \{e\})$

107. Consider the Venn diagram

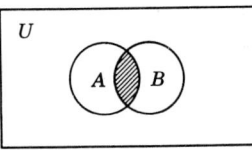

The shaded region represents ($A \subset B$ / $A \cap B$ / $A \cup B$).

$A \cup B$

108. Consider the Venn diagram

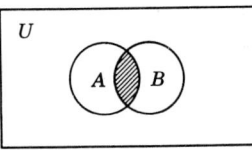

The shaded region represents [$A \cup B$ / $A \cap B$ / $(A \cap B)'$].

$A \cap B$

Preliminary Concepts

109. Consider the Venn diagram

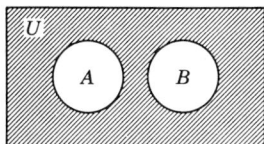

The shaded region represents $[(A \cap B)' \ / \ (A \cup B)' \ / \ A \cup B]$.

$(A \cup B)'$

Remark. We now consider another operation on sets that enables us to form a new set. However, the elements in the new set formed by this operation are unlike the elements of either of the original sets. The elements in the new set are pairs of numbers.

110. Pairs of numbers in which the order of the elements is considered to be of importance are called **ordered pairs** and are denoted by symbols such as (a, b) and (b, a). Pairing the element of $\{a\}$ with the element of $\{x\}$, we obtain two distinct _____ pairs, (a, x) and (,).

---

ordered; (x, a)

111. Write the set of ordered pairs formed by pairing elements of $A = \{a, b\}$ and $B = \{x, y\}$. List only those ordered pairs in which the elements of A are written first. $\{(\ , \ ), (\ , \ ), (\ , \ ), (\ , \ )\}$.

$\{(a, x), (a, y), (b, x), (b, y)\}$ *(Any other arrangement of the ordered pairs is satisfactory.)*

112. We pair each element of $A = \{a, b, c\}$ with each element of $B = \{x, y, z\}$ to obtain $\{(a, x), (a, y), (a, z), (b, x), (b, y), (b, z), (c, x), (c, y), (c, z)\}$. Each element such as (a, x) is called an _____ _____. There are _____ such pairs contained in this set.

---

ordered pair; nine

113. The set obtained by making matchings of the elements of the two sets A and B in the previous frame is called a **Cartesian product**. It is designated by $A \times B$, and read "the Cartesian product of A and B." If $A = \{r, s, t\}$ and $B = \{u, v\}$, then $A \times B = \{(r, u), (r, v), (s, u), (s, \ \ ), (t, u), (t, \ \ )\}$. The elements of A are listed first in each of the six _____ _____ in $A \times B$.

$(s, v); (t, v);$ ordered pairs *(Each element is a pair.)*

UNIT I

114. If $A = \{c, d\}$ and $B = \{m, n\}$, then $A \times B = \{(\ \ ,\ \ ),(\ \ ,\ \ ),(\ \ ,\ \ ),(\ \ ,\ \ )\}$.

$\{(c, m), (c, n), (d, m), (d, n)\}$ (Any arrangement of the ordered pairs is correct.)

115. $A \times B$ is called the Cartesian _____ of A and B.

product

116. If A contains 2 elements and B contains 2 elements, the Cartesian product is a set that contains _____ elements.

4

117. If $A = \{r\}$ and $B = \{s, t\}$, then $\{(r, s), (r, t)\}$ is called the _____ product of A and B. $A \times B$ is a set that contains _____ elements. Each element of $\{(r, s), (r, t)\}$ is an _____ pair.

Cartesian; two; ordered

118. If $A = \{a, b, c, d\}$ and $B = \{x, y\}$, the number of elements in $A \times B$ is _____.

8

Remark. See Exercise Ib, page 245, for additional practice on set operations and Venn diagrams.

The symbols we have used thus far are listed on the back of the response shield. You may want to refer to this list as you proceed to complete the program.

The following frames will enable you to review some additional set concepts that we have covered thus far. Note that the frame numbers of these review frames (as well as other review frames throughout the program) are preceded by the letter **R**.

R119. Any one of the things contained in a set is called a member or _____ of the set.

element

R120. Braces are used to denote a set. The elements of the set may be described by a statement or by listing the _____ within the braces.

elements (Or members.)

Preliminary Concepts

R121. $\{a, b, c, d\}$ and $\{a, c, d, b\}$ (are/are not) equal sets.

---

are *(Members do not have to be listed in any special order.)*

R122. The phrase "is a member of" is represented by the symbol _____ .

$\in$

R123. The set that contains no members is called the null set or the _____ set.

---

empty

R124. The empty set is designated by the symbol _____ .

$\emptyset$

R125. If $U = \{$mother, father$\}$, list the four subsets of U.

$\emptyset, \{$mother$\}, \{$father$\}, \{$mother, father$\}$ ($\emptyset$ *is also a subset!*)

R126. The phrase "is a subset of" is represented by the symbol _____ .

$\subset$

R127. If $A = \{a, b, c\}$ and $B = \{$mother, father, son$\}$, then the members of A can be placed in _____ _____ _____ correspondence with the members of B.

---

one-to-one

R128. If the members of two sets can be placed in one-to-one correspondence, the sets are said to be _____ .

---

equivalent

R129. Consider $A = \{a, b, c\}$ and $B = \{c, d, e\}$. Sets A and B are (equal/disjoint/equivalent).

---

equivalent

R130. If two sets, A and B, contain members such that no member of A is a member of B, then A and B are said to be _____.

---
disjoint

R131. If $A = \{a, b, c, d\}$ and $B = \{d, e, f\}$, then A and B (are/are not) disjoint.

---
are not (*d is an element of each set.*)

R132. If $T = \{\text{chair, book, table}\}$ and $W = \{\text{book, seat, pencil}\}$, then $T \cup W = \{\}$.

{chair, book, table, seat, pencil}

R133. (From Frame 132) $T \cap W = \{\}$.

{book}

R134. If the universe consists of the members of a family, that is, $U = \{\text{mother, father, son}\}$, and if $G = \{\text{son}\}$, then $G' = \{\}$.

{mother, father}

R135. The shaded region of the Venn diagram represents K ____ L.

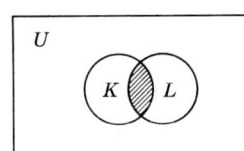

$K \cap L$

R136. The shaded region of the Venn diagram represents C ____ D.

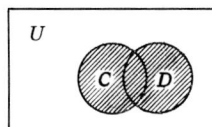

$C \cup D$

Preliminary Concepts 27

R137. The shaded region of the Venn diagram represents (C____D)'.

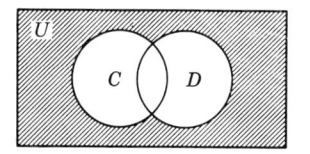

$(C \cup D)'$

R138. Each element of the set $A \times B = \{(a, x), (b, x), (a, y), (b, y)\}$ is called an _____ pair. There are _____ such pairs in this set.

---

ordered; four

R139. $A \times B$ is called the _____ product of A and B.

Cartesian

R140. Consider $A = \{c\}$ and $B = \{s, t\}$. The Cartesian product $A \times B = \{\}$.

$\{(c, s), (c, t)\}$

R141. Consider $A = \{s, t\}$ and $B = \{c\}$. The Cartesian product $A \times B = \{\}$.

$\{(s, c), (t, c)\}$ *(The members of A are listed first.)*

R142. If A contains three members and B contains four members, the Cartesian product $A \times B$ contains _____ ordered pairs.

---

twelve

Remark. Complete the following self-evaluation test on Unit I of this program and score your paper from the answers given on page 278.

UNIT I
Self-Evaluation Test

1. Bill $\in R$ is read "Bill is an ─────── of R."
2. $A \subset B$ is read "A is a ─────── of B."
3. $A \cap B$ is read "the ─────── of A and B."
4. $A \cup B$ is read "the ─────── of A and B."
5. The symbol used to represent the empty set is ───────.
6. The statement that $\{a, b\}$ is a subset of $\{a, b, c\}$ can be written symbolically $\{a, b\}$ ─────── $\{a, b, c\}$.
7. The symbol used to describe a universe is ───────.
8. The statement that A is not equal to B can be written symbolically A ─────── B.
9. $\{a, b, c\}$ and $\{c, b, a\}$ (are/are not) equal sets.
10. $\{\square, \bigcirc\}$ has (2 / 4 / 6) subsets.
11. The elements of $\{\square, \bigcirc\}$ and $\{a, b\}$ can be placed in ─── ─── ─── correspondence.
12. $\{$Mother$\}$ is (equal to/equivalent to/a subset of) $\{$Father$\}$.
13. $\{$Henry, Bill$\}$ and $\{$Henry, Paul, Judy$\}$ (are/are not) disjoint.
14. If $A = \{a, b, c\}$ and $B = \{c, d\}$, then $A \cup B$ is $\{\text{───────}\}$.
15. If $A = \{a, b, c\}$ and $B = \{c, d\}$, then $A \cap B$ is $\{\text{───────}\}$.
16. If $U = \{a, b, c\}$ and $A = \{a, b\}$, then $A' = \{\text{───────}\}$.
17. The Venn diagram corresponds to [$A \cap B$ / $A \cup B$ / $(A \cup B)'$].

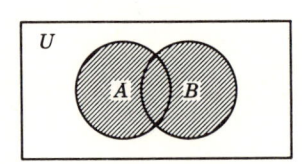

Preliminary Concepts

18. The Venn diagram corresponds
    to [$A \cap B$ / $A \cup B$ / $(A \cup B)'$].

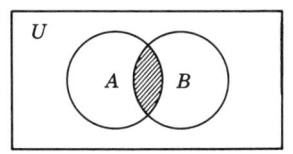

19. If $A = \{c, f\}$ and $B = \{g, e\}$, then $A \times B =$ _____.
20. If $A = \{\triangle\}$ and $B = \{\bigcirc, \square\}$, then $B \times A =$ _____.

Remark. If you missed fewer than seven questions on this test, you are ready to continue on to Unit II of this program. If you missed seven or more questions, you would probably profit in returning to the remark on page 3 and reading through the program to this point.

UNIT II

The Set of Whole Numbers

OBJECTIVES

Upon completion of this part of the program, the reader will be able to:

1. Use the correct vocabulary and/or the correct notation to express concepts associated with whole numbers.

2. Associate whole numbers with points on a number line.

3. Identify axioms of equality and order.

4. Associate the word "addition" with a binary operation on a pair of whole numbers which yields a whole number associated with the union of two disjoint sets.

5. Identify the axioms adopted for the operation of addition.

6. Associate the word "multiplication" with a binary operation on a pair of whole numbers that yields a whole number associated with the number of elements in a Cartesian product.

7. Identify the axioms adopted for the operation of multiplication.

8. Consider a theorem as being a logical consequence of a set of axioms and identify several theorems in the set of whole numbers.

9. Associate the word "subtraction" with an operation based on the process of addition.

10. Associate the word "division" with an operation based on the process of multiplication.

Remark. Let us see how some of the set concepts introduced in Unit I of this program can help us consider what is meant by the word "number," as it is used in arithmetic.

1. Recall that equivalent sets are sets whose elements can be placed in one-to-one correspondence. {horse, dog, cow} and {a, b, c} (are/are not) equivalent sets.
   ---
   are

2. {a, b} and {□, △, ○} are not equivalent. On the other hand {g} and {□} are _____ sets.
   ---
   equivalent

3. Consider all the sets that can be placed in one-to-one correspondence with {a}. The only property they share is that they are all _____ to {a}.
   ---
   equivalent

4. The property shared by all sets equivalent to a given set when all properties of the sets, except the fact of their equivalence, are completely disregarded is called **number**. The property that all sets equivalent to {a} have in common is called the number one. The property shared by all sets equivalent to {a, b} is called the number _____.
   ---
   two

5. The property that {Mary, Bill, John}, {○, △, □}, and {*, +, −} have in common is called the _____ three.
   ---
   number

6. The property that {wheels on a car} and {○, △, □, *} have in common is called the number _____.
   ---
   four

7. If A = {fingers on one hand} and B = {toes on one foot}, then A is equivalent to B. The property that A and B have in common is the _____ five.
   ---
   number

31

32   UNIT II

8. The number associated with the empty set is zero. If $A = \emptyset$, $B = \{*\}$, $C = \{*, \square\}$, and $D = \{*, \bigcirc, \square\}$, the number of B is one, the number of A is _____ , and the number of D is _____ .

---

zero; three

9. If S is equivalent to T, the property that S and T have in common when all properties of both sets, except the fact of their equivalence, are completely disregarded is called _____ .

---

number (*The members are in one-to-one correspondence.*)

10. Because $\{a, b, c, d, e, f, g\}$ and $\{m, n, o, p, q, r, s\}$ are equivalent, they are associated with the same _____ .

---

number

11. When a number is associated with the concept of "how many" members a set contains, it is being used in a **cardinal** sense. The number is called the cardinality of the set. The cardinality of $\{a, b, c, d\}$ is _____ .

---

four

12. When "five" is used to describe a property of "fingers on one hand," "five" refers to the number in a _____ sense.

---

cardinal

13. Because the number associated with the empty set is zero, the cardinality of $\emptyset$ is _____ .

0

Remark. In order to talk about the number associated with a set, the number is given a name. For example, we have already discussed the notion of "zero," "one," "two," "three," "four," and "five." Similarly, there exist sets with cardinality "six," "seven," and so on; the names are different in each case. The only difference between the notion of a number and its name and the notion of, say, a man and his name, is that, in the latter case, we can point to the man, if need be, and say, "This is John." We have nothing specific, however, to which to point when we say, for example, "This is ten."
▼

The Set of Whole Numbers

The abstract nature of number does not, however, prevent us from giving a name to a number, or representing it by a symbol, and then using the name or symbol to discuss properties we conceive to be associated with the number, although *the distinction between a number and its name or symbol should be kept clearly in mind.* Let us explore this a little further.

14. The symbol "0" represents the number zero. The symbol "1" represents the number one. The symbol "2" represents the _____ two.

   number

15. "3" is a _____ that represents the _____ three.

   symbol; number

16. Many different symbols for numbers have been used through the ages. For example, the number five has been represented by the Arabic symbol "5", the Roman symbol "V", and the tally notation "𝍬". The symbols "6", "VI", and "𝍬 /" represent the number _____ .

   six

17. Symbols are used to represent numbers; the numbers themselves are abstractions. The property that a pair of aces and twins have in common is the number two. The Arabic symbol "2" _____ the number two.

   represents (Or denotes or any similar word or phrase.)

18. Symbols such as "0", "1", "2", "3", "4", used to represent the abstract idea of a number, are called **numerals**. The numeral "5" represents the _____ five.

   number

19. "6", "7", "8", "9", etc. are examples of number symbols called _____ .

   numerals

20. Recall that a number used in a cardinal sense is associated with the concept of "how many." The number of members in $\{a, e, i, o, u, y\}$ can be represented by the numeral "___".

34   UNIT II

Remark. Notice that we have been using quotation marks around symbols such as "6" because we wish to *refer to the symbol and not to the number* it names. This is a common practice, and we shall adhere to it for awhile whenever we want to direct attention to the numeral (symbol) and not the number. When we want to talk about the number six, to which "6" refers, however, we shall write 6 rather than use the involved phrase "the number whose numeral is '6'." This is also common practice, and should cause no confusion, certainly no more than would be caused when we say "John is heavy," rather than "The man whose name is 'John' is heavy." Furthermore we shall use = and ≠ to relate numbers as well as to relate sets, with the understanding that *number symbols* displayed on opposite sides of = denote the same *number*. If the symbols are different, we can think of the symbols as being different names for the same number.

The cardinalities of all sets can, themselves, be viewed as a set, a set of numbers. Our next concern is to establish that the elements of this set can be ordered, that is, that we can arrange them in such fashion that each number has a clear and unique place in the arrangement.

21. The numbers associated with the cardinalities of nonempty sets are called **natural numbers**. 1, 2, 5, 279, 3342, and 6,874,118 are examples of _____ numbers.
    ----
    natural

22. $\{2, 5, 9\}$ is a set whose three members are _____ numbers.
    ----
    natural

23. By inserting 0 in the set of natural numbers, the set of **whole numbers** is formed. Thus, $\{0, 1, 2, 3, ...\}$ is the set of _____ numbers.
    ----
    whole

24. The set of natural numbers is a subset of the set of _____ numbers.
    ----
    whole

25. Let $A = \{*\}$ and $B = \{\triangle, \square\}$. In a one-to-one pairing of the elements of A and B, if the element * in A is paired with the element $\triangle$ in B, then there is no further element in A that can be paired with the element $\square$ in B. A has fewer members than B. The cardinality of the set with fewer members is said to be **less than** the
    ▼

The Set of Whole Numbers   35

cardinality of the other set. In the preceding example, the cardinality of _____ is less than the cardinality of _____.

A; B

---

26. The number 7 is less than the number 9 because there is a set whose cardinal number is 7 that has fewer members than some set whose cardinal number is 9. Similarly, the number 5 is _____ than the number 7.

---

less

---

27. If one set has an excess of members over a second set, the cardinality of the set with the excess of members is said to be **greater than** the cardinality of the second set. Thus, 7 is less than 9, while 7 is _____ than 5.

---

greater

---

28. The set of whole numbers is **ordered**, that is 0 is less than 1, 1 is less than 2, 2 is less than 3, 3 is less than 4, and so on. The whole number 8 is _____ than the whole number 9 and _____ than the whole number 7.

---

less; greater

---

29. The symbol $<$ is used to represent the words "is less than." $4 < 5$ is read "four is less than five." $7 < 15$ is read "seven is _____ _____ fifteen."

---

less than

---

30. Recall that the symbol = means "is the same as" but is read "is equal to" or "equals." Thus, 3 = 3 is read "three is equal to three"; 5 = 5 is read "_____ _____ _____ _____ _____."

---

five is equal to five   (*Or five equals five*)

---

31. The symbol $>$ is used to represent the words "is greater than." $5 > 3$ is read "five is greater than three." $7 > 2$ is read "_____ _____ _____ _____ _____."

---

seven is greater than two

32. 2 < 7 means the same as 7 > 2. That is, if 2 is less than 7, then 7 is _____ _____ 2.

---
greater than

33. 6 < 9 means the same as 9 _____ 6.

---
9 > 6

34. The relative order of 11 and 24 could be expressed symbolically as 11 _____ 24 or 24 _____ 11.

<; >

Remark. We have now seen how the set of natural numbers (a subset of the set of whole numbers) is comprised of all of the possible cardinalities of nonempty sets, and how these numbers can be ordered. Because of the property of order, the natural numbers can be used in another useful way.

35. Suppose the members of the set $A = \{⊟, ⊞, ⊠, ■, ⊘, ...\}$ are placed in one-to-one correspondence with the set of natural numbers $B = \{1, 2, 3, 4, 5, ...\}$ in the following way:

$$A = \{⊟, ⊞, ⊠, ■, ⊘, ...\}$$
$$\updownarrow \; \updownarrow \; \updownarrow \; \updownarrow \; \updownarrow$$
$$B = \{1, \; 2, \; 3, \; 4, \; 5, \; ...\}$$

The position of a number in A is noted by the associated number in B. The completely shaded square is in the _____ th position.

4th

36. A number used to refer to the position of a member in an ordered set is said to be used in an **ordinal** sense. When the number 2 refers to the *position* of the shaded square in $\{⊞, ■, ⊟, ⊠, ⊘,\}$ it is used in an _____ sense. When 5 is used to tell *how many* members there are in the set, it is used in a cardinal sense.

---
ordinal

The Set of Whole Numbers 37

37. If John ranks 5 in a class of 30 students, 30 is a cardinal use of the number because it tells *how many* students are in the class. 5 is an _____ use of the number because it refers to John's *position* in the set of 30 students.

ordinal

38. John works on the 3rd floor of a 5-floor office building. In this statement, 3 is used in an _____ sense and 5 is used in a _____ sense.

ordinal; cardinal

Remark. You probably have observed the rather contrived manner of using numerals while discussing ordinality in the last few frames. Normally, ordinal usage employs the words "first," "second," "third," and so on, rather than "one," "two," "three," but regardless of how such an idea is expressed, numbers form the basis for the concept of ordinality.

The standard sets we have been discussing are also important in the counting process. For this reason the natural numbers are also called **counting numbers**. Let us use the counting notion to talk about two different kinds of sets.

39. If a set is empty, or if its members can be counted, and further, if it has a last member, then it is called a **finite set**. $\{a, e, i, o, u, y\}$ is a _____ set.

finite

40. If the process of counting the elements of a set continues without end, the set is called an **infinite set**. The set of natural numbers is an _____ set.

infinite

41. Some infinite sets of numbers can be represented by using three dots within the braces. Thus, $\{1, 2, 3, \ldots\}$ represents the set of natural numbers, where the three dots mean that the set (does/does not) contain a last member.

does not

42. $\{5, 10, 15, 20\}$ is a(n) (finite/infinite) set. $\{5, 10, 15, 20, \ldots\}$ is a(n) (finite/infinite) set.

finite; infinite

UNIT II
38

Remark. Our next concern is to develop ways of discussing members of sets in general, rather than a particular member of any set. For example, while we may speak of Jane as a particular member of a class, it is often useful to speak of a "student" in the class. We are at the point where we can speak of the number 3 or the number 5; let us see how we can speak of any whole number in a given set of whole numbers.

43. A symbol, called a **variable**, is used to refer to any member of a set without specifying a particular member of the given set. Thus, if a is used to represent an unspecified (any) member of $\{1, 7, 9\}$, a is a _____.

    ---
    variable

44. Many different symbols may be used as variables. Letters of the alphabet such as a, b, c, x, y, and z are frequently used. When used to represent an unspecified member of a set, the letters are called _____.

    ---
    variables

45. A variable represents any element of a given set of numbers. The set of numbers is called the **replacement set** of the variable. Thus, if a represents a member of $\{1, 2, 3, 4\}$, then $\{1, 2, 3, 4\}$ is the _____ set of a.

    ---
    replacement

46. If a represents an element of $\{1, 2, 3, 4, 5\}$, then the replacement set of a is $\{\ _____\ \}$.

    $\{1, 2, 3, 4, 5\}$

47. If b represents an unspecified (any) element of $\{10, 15, 20, 25\}$, then b is a _____.

    ---
    variable

48. Symbols other than letters can be used as variables. For example, the symbol □ is a _____ when used to represent an element of $\{1, 7, 9\}$. The replacement set of □ is _____.

    ---
    variable; $\{1, 7, 9\}$

The Set of Whole Numbers   39

Remark. As we shall see, variables are extremely helpful in many ways. Of great importance is the fact that they enable us to make generalizations about the properties of numbers and operations on numbers. Also, variables are used very effectively to describe membership in a set especially if the set contains a large number of elements. In this case it is not practical to list all the elements. Instead, a collection of symbols called **set-builder notation** is used. From this point on, we shall name the set of whole numbers by W and the set of natural numbers by N.

49. The symbols

    $$\{a \mid a < 100, a \in W\}$$

    are read "the set of all a such that a is less than 100 and a is an element in the set of whole numbers. The symbols

    $$\{a \mid a < 50, a \in W\}$$

    are read "the set of all a _____ _____ a is less than 50 and a is an element in the set of whole numbers.
    ----
    such that

50. Note that the vertical line in the sets in the preceding frame is read "such that." Thus, $\{a \mid a \in W\}$ is read: "The set of all a _____ _____ a is an element in the set of _____ numbers.
    ----
    such that; whole

51. The symbols $\{a \mid a < 5, a \in W\}$ can be written by listing the members as $\{0, 1, 2, 3, 4\}$. Similarly, $\{a \mid a < 3, a \in W\}$ can be written by listing the members as $\{\}$.

    $\{0, 1, 2\}$

52. The symbols $\{a \mid a > 7, a \in N\}$ are read "the set of all a _____
    _____."
    ----
    such that a is greater than 7, and a is an element in the set of natural numbers.

53. List the members of the set $\{a \mid a > 7, a \in N\}$.

    $\{8, 9, 10, \ldots\}$

40 UNIT II

54. The set in the preceding frame is (finite/infinite).

infinite

55. The symbols $\{a \mid 2 < a < 7, a \in W\}$ are read: "The set of all a such that 2 is less than a, a is less than 7, and a is an element in the set of whole numbers." The symbols $\{a \mid 1 < a < 5, a \in W\}$ are read: "The set of all a such that ___ ___ _____ ___, ___ _____ _____ ___, and a is an element in the set of whole numbers."

1 is less than a, a is less than 5

56. The first set in the preceding frame can be written by listing its members as $\{3, 4, 5, 6\}$. The second set can be written by listing its members as $\{\}$.

$\{2, 3, 4\}$

Remark. Our next concern will be to associate the set of whole numbers with points on a line. The following figure, which shows the set of natural numbers associated with the right-hand end points of line segments of equal length, is called a **number line**.

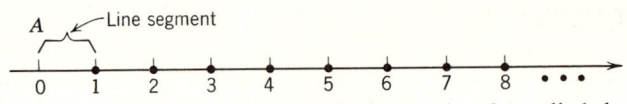

The point at the letter "A" and associated with the number 0 is called the **origin**.

Of any two natural numbers associated with points on the number line, the greater number is associated with a point to the right of that associated with the smaller. For example, the point associated with the number 8 is to the right of the point associated with the number 7. An arrowhead is drawn on the right of the number line to show this orientation.

57. Numbers assigned to selected reference points are called **scale numbers**. The point labeled A on the number line

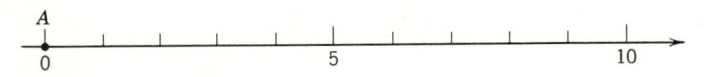

is the origin and 0, 5, and 10 are _____ numbers.

scale

The Set of Whole Numbers 41

Remark. A scale number simply tells you how many equal segments have been laid off from the origin on the number line. When the number 1 has been assigned to the right-hand end point of a line segment, and segments equal to this segment have been marked off along the number line, we refer to each of these equal segments as a *unit segment*, or, sometimes, just a *unit*. Therefore, in the preceding frame, the scale numbers 5 and 10 below the number line tell us that from the origin to the point representing 5 we have laid off exactly 5 units. On the other hand, the number line

```
   |  |  |  |  |  |  |  |  |  |
  100                         200
```

has scale numbers that tell us that each small mark along the line represents 10 unit segments or simply 10 units.

If a number line is extended sufficiently far to the right, or if the scale is chosen appropriately, *any given element of the set of whole numbers can be associated with a unique (one and only one) position on the number line.*

58. Marking the position on the line associated with a number is called **graphing** the number. A dot is usually used to mark the position. Thus

```
   |  |  |  •  |  |
  40           50
```

shows the graph of the number _____. Each segment on the graph represents _____ units.

46; 2

59. On the number line

```
  A     B
  •  |  •  |  |  |  |  |  |  |  |  |  |  |  |  |
  0        5           10          15
```

the point labeled A is the _____ and the point labeled B is the _____ of the number 3. The indicated scale numbers are 0, 5, 10, and _____.

---

origin; graph; 15

60. The _____ of the numbers 6 and 9 are shown on the number line

```
   |  |  •  |  |  •  |  |
        5          10
```

5 and 10 are _____ numbers.

---

graphs; scale

61. The number corresponding to a point on a number line is the **coordinate** of the point. The point marked on the number line

has a coordinate of _____. The point is called the _____ of the number.

26; graph

62. On the number line

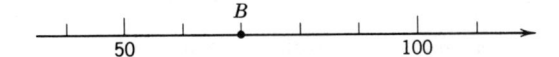

70 is the _____ of the point labeled B. The point is the _____ of 70.

---

coordinate; graph

63. Scale the given line segment and then graph $\{6, 7, 9\}$.

Label these points above the line with the numerals 6, 7, and 9.

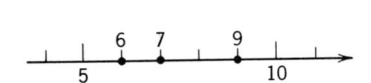

(Perhaps you didn't use the same scale numbers we used; there are many possibilities.)

64. A number line shows very clearly whether one number is less than, equal to, or greater than another. Of any two numbers, the point associated with the *smaller* always lies to the *left* of the point associated with the larger (if the arrowhead is on the right). Without knowing the scale on the number line

we can say that the number associated with point A is (less than/greater than) the number associated with point B.

---

less than

The Set of Whole Numbers 43

65. On the number line

```
                    C         D
    ─────────────┼─────────┼────────►
```

the number associated with the point *D* is (greater than/less than/equal to) the number associated with the point *C*.

------
greater than

Remark. This is a good place to stop and review.

R66. The property that two equivalent sets have in common, when all other properties of both sets, except the fact of their equivalence, are disregarded, is called _____ .

------
number

R67. Symbols such as "0", "1", "2", used to represent numbers, are called _____ .

------
numerals

R68. The number of the set $\emptyset$ is _____ .

0

R69. The symbol $<$ is used to represent the words "is (less than/greater than)."

------
less than

R70. The relative order of 17 and 21 could be expressed by symbols as 17 ____ 21 or 21 ____ 17.

$<$; $>$

R71. When the number 4 is associated with the concept of "how many," the number is being used in a _____ sense.

------
cardinal

R72. When the number 2 is associated with the position of the circle in $\{*, \bigcirc, \triangle, \square\}$, the number 2 is being used in an _____ sense.

ordinal

R73. If a set is empty or if the elements of a set can be counted and if the set has a last member, then it is called a _____ set. If the members in the set cannot be counted, or if a counting will never produce a last member, then it is called an _____ set.

finite; infinite

R74. $\{\text{whole numbers}\}$ is a(n) (finite/infinite) set.

infinite

R75. If $U = \{\text{natural numbers}\}$ and if $A = \{1, 2, 3\}$, then A' is a(n) (finite/infinite) set.

infinite $(A' = \{4, 5, 6, \ldots\})$

R76. A symbol used to represent an unspecified member of a given set is called a _____.

variable

R77. If $a \in \{\text{whole numbers}\}$, then the variable a represents an unspecified member of the set of _____ _____.

whole numbers

R78. A variable represents any element of a given set of numbers. The set of numbers is called the _____ set of the variable.

replacement

R79. A collection of symbols such as $\{a \mid a < 50, a \in W\}$ is called _____ notation, and is read "the set of all a _____ _____ _____ _____ _____, and a is an element in the set of whole numbers.

set-builder; such that a is less than 50

The Set of Whole Numbers   45

R80. Because the whole numbers are ordered, they can be associated with points on a _____ .

---

line (*Or number line*)

R81. Numbers are associated with points on a line by marking off equal line segments starting from an arbitrary point called the _____ .

---

origin

R82. The number corresponding to a point on the number line is called the _____ of the point.

---

coordinate

R83. The point on a number line corresponding to a number is called the _____ of the number.

---

graph

R84. If the arrowhead on a number line is on the right, the point associated with the smaller of two numbers lies to the _____ of the point associated with the larger.

---

left

Remark. See Exercise IIa, page 247, for additional practice on topics that have been considered to this point in Part II of the program.

We now have at our disposal an ordered set of numbers, the whole numbers. About all we can do with these at present is to use them to count members of a set and to establish the relative position of a member in a set. We shall, however, introduce two operations, addition and multiplication, which are performed on whole numbers and which will render the set much more useful. First we want to make some agreements and assumptions about the behavior of the numbers in our set. We can then begin to examine just what the assumptions we have made imply for these numbers.

In the frames that follow we shall adopt the customary convention of referring to variables such as *a, b,* and *c* as *whole numbers* rather than as *representing* whole numbers.

▼

UNIT II

Because we do not wish to have the same symbol used for different purposes in the same discussion, let us make the following agreement:

For all $a \in W$, $a = a$.

This is called the **Reflexive law of equality**.

85. Writing $b = b$ is a statement of the _____ law of _____. It is simply an agreement that b will always represent the same _____ in the mathematical discussion in which it is being used.

   Reflexive; equality; number

86. The Reflexive law of equality asserts that $c =$ ___ in any discussion.

   $c = c$

Remark. Let us agree, also, that if two different symbols are used for the same number, then the order in which we write a statement of equality doesn't matter. That is:

For all $a,b \in W$, if $a = b$, then $b = a$.

This is called the **Symmetric law of equality**.

87. If $c,d \in W$, then rewriting $c = d$ as $d = c$ is justified by the _____ ___ of equality.

   Symmetric law

88. "For all $h \in W$, $h = h$" is a statement of the _____ law of equality and "If $h = k$, then $k = h$" is a statement of the _____ law of equality.

   Reflexive; Symmetric

Remark. As another agreement, let us assume that:

For all $a,b,c \in W$, if $a = b$ and $b = c$, then $a = c$.

This is called the **Transitive law of equality**.

89. For all $d,e,f \in W$, if $d = e$ and $e = f$, then writing $d = f$ can be justified by the _____ ___ of equality.

   Transitive law

The Set of Whole Numbers 47

90. The Transitive law of equality asserts that for all $r,s,t \in W$, if $r = s$ and $s = t$, then ____ = ____ .

$r = t$

91. The three laws of equality we have thus far stated are the _____ law, symbolized by "$a = a$," the _____ law, symbolized by "If $a = b$, then $b = a$," and the _____ law, symbolized by "If $a = b$ and $b = c$, then $a = c$."

Reflexive; Symmetric; Transitive

Remark. These laws are simply *agreements* that we will not change the meaning of a symbol in the middle of a discussion. They are sometimes referred to as the properties of equality.

92. When an *assumption* is made in mathematics, it is customary to call the assumption an **axiom** or a **postulate**. The Symmetric law of equality is an assumption. ($\{0, 1, 2, 3, \ldots\}$ is an infinite set and it is not possible to check the fact that if $a = b$, then $b = a$ for each element in the set.) Because the Symmetric law of equality is an assumption, it is an axiom or a _____ .

---
postulate

93. An axiom or a postulate is an _____ .

---
assumption

94. The words postulate and _____ are synonomous.

---
axiom   (*Or assumption*)

95. The Reflexive law and the Transitive law for equality are assumptions; therefore they are _____ .

---
axioms   (*Or postulates*)

Remark. Another assumption that we make is stated as follows:

For all $a,b \in W$, if $a = b$, then a may be substituted for b and b for a in any expression without changing the truth or falsity of the statement.

For our purposes, we shall take this statement as an axiom and shall call it the **Substitution law of equality**. We shall see that this statement which includes the Transitive law of equality as a special case, is very useful in rewriting mathematical expressions.

In general, the terms "postulate" and "axiom" can be used interchangeably. We have elected to use the term "axiom" throughout the rest of this program. We now have four laws to govern our work with equalities: the Reflexive, Symmetric, Transitive, and Substitution laws. These laws are *agreements* or *assumptions* we make concerning the equality relation and are examples of axioms. There is another relation existing between two whole numbers, that of inequality. While we shall need to make some agreements about the use of symbolism in connection with inequalities, there is a subtle conceptual difference between the agreements about equalities and those about inequalities. When we make *statements about equality*, we are essentially talking about *just one number*. For example, $a = b$ simply asserts that *a* and *b* are different names for the same thing. *With an inequality*, however, *two distinct numbers* are involved, and agreements about the symbolism used is somewhat different than those of equality.

Now, let us make the following assumption concerning the order of whole numbers.

For all $a,b \in W$, exactly one of the following relations holds:
$a < b, \quad a = b, \quad \text{or } b < a.$

This assumption is called the **Trichotomy law**.

96. This law asserts that for any two whole numbers, one of the numbers is either greater than, equal to, or _____ than the other number.
    ------
    less

97. Since the Trichotomy law is an assumption concerning the order of whole numbers, it may be called an _____ of order.
    ------
    axiom

98. The assumption that of any two whole numbers one of the numbers is greater than, equal to, or less than the other is called the _____ law.

    Trichotomy

The Set of Whole Numbers

Remark. A second axiom of order asserts:

For all a,b,c ∈ W, if a < b and b < c, then a < c.

This axiom is called the **Transitive law of inequality**.

99. The Transitive law of inequality asserts that if one number is less than a second, and the second is less than a third, then the first is _____ _____ the third.

---
less than

100. "If $a < 7$ and $7 < b$, then $a < b$" is an example of the _____ law of inequality.

Transitive

Remark. We have now adopted the following axioms for equality:

For all $a,b,c \in W$:
- $a = a$.   Reflexive law
- If $a = b$, then $b = a$.   Symmetric law
- If $a = b$ and $b = c$, then $a = c$.   Transitive law
- If $a = b$, then a may be substituted for b Substitution law
  and b for a in any expression without
  changing the truth or falsity of the statement.

We have also adopted the following axioms for order.

For all $a,b,c, \in W$:
- Exactly one of the relations holds:
  $a < b, a = b,$ or $a > b$.   Trichotomy law
- If $a < b$ and $b < c$, then $a < c$.   Transitive law of inequality

The next sequence of frames will give you an opportunity to review the terms and concepts that are concerned with equality and inequality.

R101. An axiom is an agreement or an _____.

---
assumption (*Postulate is also correct, but it does not fit.*)

R102. Axioms are sometimes called laws. Thus, the statement "For all $a,b \in W$, if $a = b$, then $b = a$" is called the
   a. Reflexive law of equality.
   b. Symmetric law of equality.

▼

c. Transitive law of equality.
d. Substitution law of equality.

---

b (*Or Symmetric law of equality.*)

R103. The Symmetric law of equality justifies writing the equality $a = 6$ as $6 = $ _____ .

$6 = a$

R104. The statement

"For all $a,b,c \in W$, if $a = b$ and $b = c$, then $a = c$"

is called the _____ _____ of equality.

Transitive law

R105. If $a = b$ and $b = 3$, the Transitive law of equality justifies writing _____ .

$a = 3$

R106. The statement

"For all $a,b \in W$, if $a = b$, then a may be substituted for b and b for a in any expression, without changing the truth or falsity of the statement"

is called the _____ _____ .

Substitution law

R107. The statement

"For all $a,b \in W$, exactly one of the relations holds:
$a < b,\ a = b,\ a > b$."

is called the
    a. Trichotomy law.   b. Transitive law of inequality.

---

a.

R108. "If $a < b$ and $b < c$, then $a < c$" is called the _____ law of inequality.

Transitive

The Set of Whole Numbers

R109. If $b < d$ and $d < 3$, the Transitive law of inequality justifies writing:

   a. $3 < d$ b. $d < b$ c. $b < 3$

---

c.

Remark. We observe again that axioms are sometimes called laws and we use this latter term when appropriate.

See Exercise IIb, page 249, for additional practice on ideas of equality and order.

Now we shall consider *the operation of addition which pairs two whole numbers with a unique (one and only one) third number* and shall consider how this pairing is accomplished by means of sets.

110. Recall that two sets, A and B, are disjoint if they contain no common elements ($A \cap B = \emptyset$). If $A = \{a, b, c\}$ and $B = \{d, e, f, g\}$, then A and B (are/are not) disjoint.

---

are

111. Recall that the union of two sets, A and B, is the set whose members are all those members that are in A or B or both. If $A = \{a, b, c\}$ and $B = \{d, e, f, g\}$, then $A \cup B = \{$ _____ $\}$.

$\{a, b, c, d, e, f, g\}$

Remark. We can look at the **sum** of two whole numbers as that number that is the cardinality of the set formed by the *union* of two *disjoint* sets.

112. If $A = \{a, b, c\}$ and $B = \{d, e, f, g\}$, then $A \cup B = \{a, b, c, d, e, f, g\}$ and $A \cap B = \emptyset$. The cardinality of A is 3, the cardinality of B is 4, and the cardinality of $A \cup B$ is called the _____ of 3 and 4.

---

sum

113. If R and S are disjoint, then the number associated with $R \cup S$ is called the _____ of the numbers associated with R and S.

---

sum

114. If $A = \{$Mary, John, Bill$\}$ and $B = \{$Joan, Richard$\}$, then the cardinality of A is 3, the cardinality of B is 2, and the cardinality of $A \cup B$ is _____, and is called the sum of _____ and _____.

5; 3; 2

Remark. Note that addition is a binary operation associated with numbers and union is a binary operation associated with sets. The union of two sets A and B is represented by $A \cup B$. The sum of two numbers a and b is represented by $a + b$.

We shall now make certain assumptions concerning the operation of addition which are part of the basic building blocks of the system of whole numbers.

115. Both 7 and 8 are whole numbers. Their sum, $7 + 8$, is also a _____ number.

---

whole *(Of course their sum is also a natural number.)*

116. 297,314 and 6,284,121 are whole numbers. Their sum (is/is not) a whole number.

---

is

117. It seems reasonable to assume that the sum of any two whole numbers is also a _____ number.

---

whole

118. When the result of performing an operation on two numbers of a given set is always another element of the same set, the set is said to be **closed** with respect to that operation. Since it is assumed that the sum of any two whole numbers is a _____ _____, it can then be stated that the set of whole numbers is _____ with respect to the operation of addition.

---

whole number; closed

Remark. The statement:

"For all $a, b \in W$, $a + b$ is a whole number"

is called the **Closure law for addition.**

The Set of Whole Numbers

119. Because 1 is a whole number, the _____ law for addition guarantees that for any whole number a, $a + 1$ is a _____ number.

Closure; whole

120. The sum of 5 and 3 equals the sum of 3 and 5. That is, $5 + 3 = 3 + 5$. Similarly, $8 + 6 = 6 +$ ____ .

$6 + 8$

Remark. It is assumed that the sum of any two whole numbers a and b is the same as the sum of b and a. Thus:

$$\text{For all } a, b \in W, \ a + b = b + a.$$

This assumption is called the **Commutative law of addition**. The word "commutative" derives from the fact that the order of the numbers is changed, or "commuted."

121. Because the Commutative law of addition is an assumption, it is an _____ .

---

axiom

122. The _____ law of addition asserts that $2 + 3 = 3 +$ ____ .

Commutative; $3 + 2$

123. The addition operation pairs *two* whole numbers with a unique third whole number. The symbols ( ), called **parentheses**, or [ ], called **brackets**, are used to show that two numbers are to be considered grouped for the purpose of addition. Thus, $(5 + 4) + 3$ means the sum of $5 + 4$ and ____ . $5 + (4 + 3)$ means the sum of 5 and ____ .

3; $4 + 3$ (or 7)

Remark. It is assumed that:

$$\text{For all } a, b, c \in W, \ (a + b) + c = a + (b + c).$$

This assumption is called the **Associative law of addition**.

124. The Associative law of addition is concerned with the grouping, or "associating," of terms in a sum. Since this law is assumed, it is an _____.

axiom

125. Addition is a **binary operation**. That is, to add the members of any given collection of numbers, the operation of addition is performed on only _____ numbers at a time.

two

Remark. We have now defined a collection of symbols such as 2 + 5. What about a collection of symbols such as 2 + 5 + 3? We shall define $a + b + c$ to mean the sum of $a + b$ and c. Thus:

For all $a, b, c \in W$, $a + b + c = (a + b) + c$.

126. The fact that 2 + 5 + 3 can be viewed either as (2 + 5) + 3 or as 2 + (5 + 3) is an application of the _____ law of addition.

Associative

127. The Commutative law of addition is concerned with the *order* of terms. The Associative law of addition is concerned with the *grouping* of terms. When $a + 2$ is written $2 + a$, only the _____ of the terms has been changed. When $(c + d) + 2$ is written $c + (d + 2)$, only the _____ of the terms has been changed.

order; grouping

128. The law that permits writing $3 + b$ as $b + 3$ is the _____ law of addition.

Commutative *(Only the order is changed.)*

129. The law that permits writing $(2 + c) + d$ as $2 + (c + d)$ is the _____ law of addition.

Associative *(Only the grouping has been changed.)*

The Set of Whole Numbers 55

130. If $a = 3$, b is a whole number, and $a + b = 5$, the Substitution law of equality justifies writing $3 + b = 5$. Similarly, if $n = 4$, m is a whole number, and $m + n = 6$, then $m +$ ____ $= 6$.

$m + 4 = 6$

131. If r and s are whole numbers, and if $r = s$ and $r + t = 12$, then the _____ law justifies writing $s + t = 12$.

Substitution   (*s has been substituted for r.*)

Remark. There is a close relationship between "substitution" and "replacement." Thus, if a, b, c, and d are whole numbers, and if $a = b$, then if we wish to write $b + c = d$ in place of $a + c = d$, one way to look at it is that we have simply replaced a with b in the statement $a + c = d$.

132. If $a = 9$ and $a + b = 4$, then when we invoke the Substitution law of equality to write $9 + b = 4$, we can look at the procedure as a replacement of ____ by ____.

---
a; 9

Remark. We have observed that the number 0 was the cardinality of the empty, or null, set. Hence, the number 0 has an important property. We shall take as an axiom the following property of 0 in the set of whole numbers:

For all $a \in W$, $a + 0 = a$ and $0 + a = a$.

This statement is called the **Identity law of addition**.

133. For example, using this law, $7 + 0 = 7$, $0 + 207 =$ _____, and $18 +$ ____ $= 18$.

207; 0

134. When an operation on two members of a set produces one of the two members, the other member is called the **identity element** for that operation. Thus, as used in the previous frame, 0 is the _____ element for addition in the set of whole numbers.

---
identity

Remark. You have seen that the operation of addition on whole numbers is a binary operation that pairs two whole numbers with a unique third number. This operation is governed by four basic axioms:

For all $a,b,c \in W$:	
$a + b \in W$	Closure law
$a + b = b + a$	Commutative law
$(a + b) + c = a + (b + c)$	Associative law
$a + 0 = a$ and $0 + a = a$	Identity law

These axioms, together with our axioms of equality, justify making other assertions about whole numbers. Such assertions, that are consequences of our assumptions, are called **theorems**. *A demonstration of the truth of a theorem is called a proof.* While it is not an objective of this program to teach you to make proofs, it is desirable that you be able to follow a logical argument, and, from time to time hereafter, we will make such arguments in detail. For example, here is a theorem:

$$\text{If } a,b,c \in W, \text{ and if } a = b, \text{ then } a + c = b + c.$$

Notice that this theorem consists of *two parts*. First, we are given some things as being true, namely, that a, b, and c are whole numbers, and that $a = b$. Second, the assertion is made that these given things imply that $a + c = b + c$. *The portion of a theorem that gives us information as being true is called the **hypothesis**, and the portion of a theorem that follows from the given information is called the **conclusion**.* Let us see if we can use our axioms to argue that the conclusion in the above theorem follows logically from the hypothesis.

135. We know that a, b, and c are whole numbers because it is given information (i.e., it is a part of the hypothesis of the theorem). We know that $a = b$ because it is also _____ _____ .

---
given information

136. Since a and c are whole numbers, we know that $a + c$ is a whole number also, because the Closure axiom for addition states that the sum of any two whole numbers is a whole number. Furthermore the Reflexive law of equality asserts that $a + c = a +$ ___ .

$a + c$

The Set of Whole Numbers

137. Bur, since $a = b$, and $a + c = a + c$, we can write $a + c = b + c$, where a has been replaced with b on the right side of the statement. This is justified by the _____ law.

Substitution

Remark. That is all there is to it. This is what we wanted to prove; that is, if $a = b$, then $a + c = b + c$. Of course, this is an extremely brief and simple argument, but the result is useful for later work.

It is not essential that you be able to construct such a proof yourself. It is, however, desirable that you be able to follow the argument.

We now show the same proof in another form, a concise statement–reason format.

Theorem: If $a, b, c \in W$, and if $a = b$, then $a + c = b + c$.

Proof:

$a = b$	(given information)
$a + c \in W$	(Closure law for addition)
$a + c = a + c$	(Reflexive law of equality)
$a + c = b + c$	(Substitution law of equality)

Since we have demonstrated the validity of the theorem for any whole numbers a, b, and c, we can use it to justify writing things like $r + 6 = s + 6$, if we are given that r and s are whole numbers and $r = s$.

138. The theorem we proved above asserts that:

$$\text{If } a, b, c \in W, \text{ and if } a = b, \text{ then } a + c = b + c.$$

Thus, if $a = b$, then $a + 5 = b + 5$ and $a + 9 = $ _____.

$b + 9$

139. If $r = s$, the theorem we have established justifies writing $\underline{r + 3 = s + }$ _____.

$r + 3 = s + 3$

140. Statements that are assumed to be true are called _____ ; statements that follow logically from assumptions are called _____ .

---
axioms; theorems

58 UNIT II

Remark. We may write 3 + 8, or 8 + 3, to denote the sum of 8 and 3, or we may simply write 11. Since a number can be named in many ways, it would be convenient if we could designate one of these as basic, and then, when we wish to use this basic name, we can refer to the *basic numeral* for the number. The numeral we shall designate as basic is the one that is in the set {"0", "1", "2", "3", ...}. We shall employ it without the quotation marks, however.

141. 8 + 6 is a name of a number. The basic numeral for 8 + 6 is _____.

14

142. The sum 7 + 5 can be written as the basic _____ 12.

------

numeral

143. The sum, 3 + 8 or 8 + 3, can be written as the _____ _____ 11.

------

basic numeral

Remark. It's true that if we wished to find the basic numeral associated with the union of two disjoint sets of concrete objects, we could combine the sets and count the members. If the objects are too large to push around, we could represent these on our fingers or by some other physical means and determine the number associated with the union of these two sets. As a start in getting around such an unsophisticated process, we become familiar with certain basic facts of addition, facts that can be set down in the form of a table, namely, the basic numerals for a number of expressions of the form $a + b$. Note that a basic numeral may correspond to a number of different sums, although any single sum corresponds to just one basic numeral.

So far we have been able to obtain a numeral for a sum only by counting the elements in the set formed by the union of two disjoint sets. Let us assume that we have now done this for the sums of numbers 1 through 10 and summarized this information in a compact form in the following addition table.

The Set of Whole Numbers 59

+	1	2	3	4	5	6	7	8	9	10
1	2									
2	3	4								
3	4	5	6							
4	5	6	7	8						
5	6	7	8	9	10					
6	7	8	9	10	11	12				
7	8	9	10	11	12	13	14			
8	9	10	11	12	13	14	15	16		
9	10	11	12	13	14	15	16	17	18	
10	11	12	13	14	15	16	17	18	19	20

Note that we can use an abbreviated addition table because we are familiar with the commutative law of addition. Once we know that two plus seven equals nine, the commutative law assures us that seven plus two also equals nine. Furthermore, from the Identity law of addition, we know that for all $a \in W$, $a + 0 = a$ and $0 + a = a$.

How about the sums of numbers greater than ten? Is it necessary for us to memorize these sums also? The answer is no, and the next sequence of frames will illustrate the applications of the Associative law of addition to the familiar concept of "regrouping" in addition. However, because we do not wish to wander off into a discussion of the properties of our numeration system, we shall simply assume that if we know that $3 + 5 = 8$, then we know also that $30 + 50 = 80$, $300 + 500 = 800$, and so on.

144. 11 can be written $10 + 1$; 24 can be written $20 +$ ___ .

$20 + 4$

145. The sum of 24 and 8 can be written as $(20 + 4) + 8$ or as $20 + (4 + 8)$, by applying the _____ law of addition.

Associative

146. From our table of addition facts (or, we hope, from experience), the sum of 4 and 8 can be written as 12 and then as $10 + 2$. Thus, $20 + (4 + 8) = 20 + (10 +$ ___ $)$.

$20 + (10 + 2)$

147. By the Associative law of addition, 20 + (10 + 2) can be written as (20 + ___) + ___ , which leads to 30 + 2 or 32.

(20 + 10) + 2

148. A complete representation for finding the sum of 24 and 8 is given by the following.

$$\begin{aligned}
\text{Step 1.} \quad & 24 + 8 = (20 + 4) + 8 \\
\text{Step 2.} \quad & = 20 + (4 + 8) \\
\text{Step 3.} \quad & = 20 + 12 \\
\text{Step 4.} \quad & = 20 + (10 + 2) \\
\text{Step 5.} \quad & = (20 + 10) + 2 \\
\text{Step 6.} \quad & = 30 + 2 \\
\text{Step 7.} \quad & = 32
\end{aligned}$$

The Associative law of addition justifies our regrouping numbers and writing step 2 from step 1 and step ___ from step ___ .

5; 4

Remark. The purpose of the previous frame is to show that the *addition algorithm* (a procedure for computation), in which numerals are regrouped, is really the application of properties of whole numbers and our numeration system. The seven steps constitute the statements of a proof; however, the justifications for moving from one step to the next have not been stated. Note that, in step 3, 4 + 8 is rewritten as 12. Then, 12 is expressed as 10 + 2 and the Associative law is invoked in step 5 to rewrite 20 + (10 + 2) as (20 + 10) + 2. This is the rationale underlying familiar statements such as "To add 24 and 8, find the sum of 4 and 8, write down the '2' and 'carry 1' to the next column."

The following frames will give you an opportunity to review some of the fundamental properties related to the operation of addition on the whole numbers.

R149. If A and B are disjoint, then the number associated with $A \cup B$, the union of A and B, can be viewed as the ___ of the numbers associated with A and B.

---

sum

R150. When the result of adding two elements of a given set is always another element in the same set, the set is said to be (closed/commutative/associative) with respect to addition.

---

closed

The Set of Whole Numbers 61

R151. When the result of applying an operation on two numbers is the same, regardless of the order in which the numbers are taken, then the operation is said to be (closed/commutative/associative).

commutative

R152. When the result of applying two operations on three numbers is the same, regardless of the way the numbers are grouped, then the operations are said to be (closed/commutative/associative).

associative

R153. In mathematics, a formal assumption is called a(n) _____.

axiom *(Or postulate.)*

R154. In addition operations, the Closure law, the Commutative law, and the Associative law are assumed. Therefore, these laws are _____.

axioms

R155. The assumption that the addition of two whole numbers is always another whole number is called the _____ law for addition.

Closure

R156. The _____ law of addition justifies writing 3 + 7 as 7 + 3.

Commutative

R157. The _____ law of addition justifies writing (10 + 3) + 6 as 10 + (3 + 6).

Associative

R158. The _____ law of addition justifies writing 5 + 0 = 5 and 0 + 5 = 5.

Identity

62   UNIT II

R159. The number 0 is the _____ _____ for addition in the set of whole numbers.

---
identity element

R160. Assertions that follow logically from definitions and axioms are called _____ .

---
theorems

R161. Complete the following theorem:

If $a, b, c \in W$, and if $a = b$, then $a + c =$ _____ $+$ _____ .

$a + c = b + c$

Remark. See Exercise IIc, page 250, for additional practice on the operation of addition.

Let us turn next to the operation of multiplication on whole numbers and the assumptions we make about this operation.

Like addition, multiplication is also a binary operation and is denoted by the symbol $\times$ or $\cdot$. The operation pairs with each two whole numbers, a unique whole number called their **product**.

162. Thus, 4×5 or $4 \cdot 5$ is the _____ of 4 and 5.

---
product

163. The numbers that are paired to yield a product are called the **factors** of the product. For example, 82 and 341 are factors of the _____ $82 \cdot 341$.

---
product

164. 2 and 3 are _____ of the _____ $2 \cdot 3$.

---
factors; product

165. For reasons that we shall see later, it is convenient at times to place one or both factors of a product in parentheses. $5 \cdot (7)$ means "5 times 7." $6 \cdot 8, 6 \cdot (8)$,

▼

The Set of Whole Numbers 63

and 6 × 8 all mean "6 _____ 8."

---

times

166. 9 · 11 and 9 · (11) both indicate the product of the _____ 9 and 11.

---

factors (*Or numbers*)

167. (*a*) · (*b*) means the same as *a* · (*b*) or *a* · *b*. Each of these represents the _____ of the _____ *a* and *b*.

---

product; factors

Remark. The product of two whole numbers can be viewed in two ways. Perhaps you have considered a product as a number that could be obtained as the sum of successive additions. Thus, 4 · 5 would represent the sum 5 + 5 + 5 + 5. We shall adopt a more abstract approach but one that is consistent with the concept of successive additions and has the advantage that the approach can be carried over to other areas of mathematics.

We previously used sets to interpret sums. Products can also be interpreted through the use of sets.

168. Recall that pairs of numbers in which the order of the elements is considered to be of importance are called ordered pairs and are denoted by symbols such as (3, 1), (1, 3), (*a*, *b*), and (*b*, *a*). Pairing the element of {*a*} with the element of {*x*}, we obtain two distinct _____ pairs, (*a*, *x*) and (,).

---

ordered; (*x*, *a*)

169. Write the set of ordered pairs formed by pairing elements of *A* = {*a*, *b*} and *B* = {*x*, *y*}. List only those ordered pairs in which the elements of *A* are written first. {(,), (,), (,), (,)}.

{(*a*, *x*), (*a*, *y*), (*b*, *x*), (*b*, *y*)} (*Any other arrangement of the ordered pairs is satisfactory.*)

170. We pair each element of $A = \{a, b, c\}$ with each element of $B = \{x, y, z\}$ to obtain $\{(a, x), (a, y), (a, z), (b, x), (b, y), (b, z), (c, x), (c, y), (c, z)\}$. Each element such as (a, x) is called an _____ _____. There are _____ such pairs contained in this set.

---

ordered pair; nine

Remark. Recall that the set obtained by making matchings of the elements of the two sets A and B in the previous frame is called a Cartesian product. It is designated by $A \times B$, and read "the Cartesian product of A and B."

171. If $A = \{r, s, t\}$ and $B = \{u, v\}$, $A \times B = \{(r, u), (r, v), (s, u), (s, \_\_), (t, u), (t, \_\_)\}$. The elements of A are listed first in each of the six _____ _____ in $A \times B$.

(s, v); (t, v); ordered pairs   *(Each element of $A \times B$ is a pair.)*

172. If $A = \{c, d\}$ and $B = \{m, n\}$, $A \times B = \{(\_\_, \_\_), (\_\_, \_\_), (\_\_, \_\_), (\_\_, \_\_)\}$.

$\{(c, m), (c, n), (d, m), (d, n)\}$ *(Any arrangement of the ordered pairings is correct.)*

173. $A \times B$ is called the Cartesian _____ of A and B.

---

product

174. If A contains 2 elements and B contains 2 elements the Cartesian product is a set that contains _____ elements.

4

175. If $A = \{r\}$ and $B = \{s, t\}$, then $\{(r, s), (r, t)\}$ is called the _____ product of A and B. $A \times B$ is a set that contains _____ elements. Each element of $\{(r, s), (r, t)\}$ is an _____ pair.

Cartesian; two; ordered

The Set of Whole Numbers

176. If $A = \{a,b,c,d\}$, then the number of elements in set A is 4. If $B = \{x,y\}$, the number of elements in set B is _____. The number of elements in $A \times B$ is $4 \cdot$ _____.

2; 4 · 2

Remark. The previous frames suggest that the product of two whole numbers can be interpreted through a Cartesian product. The following frames suggest another way to interpret a product.

177. A **lattice**, or an **array**, is an orderly arrangement of objects (called **elements**) in rows (horizontal) and columns (vertical). For example, a lattice of 4 columns and 3 rows looks like this:

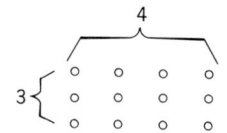

The lattice

contains _____ columns and _____ rows. By counting, we can determine that this lattice has (6 / 12 / 24) elements.

6; 2; 12

178. Draw a lattice that has 2 columns and 3 rows.

○ ○
○ ○
○ ○

179. By counting we can determine that there are 6 _____ in the lattice in Frame 178.

---

elements *(Or members, or objects.)*

180. A product can be thought of as a number associated with the elements in a lattice where the number of rows and the number of columns are factors of the product. Consider a lattice of 3 columns and 4 rows.

▼

By counting we determine that the lattice contains _____ elements.

12

181. The lattice in Frame 180 is a picture of the statement 3 · 4, where 3 denotes the number of columns in the lattice and 4 the number of _____.

---

rows

182. The lattice

○ ○ ○ ○ ○ ○
○ ○ ○ ○ ○ ○
○ ○ ○ ○ ○ ○

is a picture of the statement _____.

---

6 · 3

183. The lattice showing the product of 7 and 3 would have _____ elements.

21

Remark. Of course, you didn't draw a lattice and count the elements. You know your multiplication table. If you didn't, you would have to count the elements in the lattice.

We have now observed that the number associated with a product can be visualized either through a Cartesian product or through an array. In fact, an array or a lattice is simply the **graph** of a Cartesian product.

184. If $A = \{a, b, c\}$ and $B = \{d, e, f, g\}$, the Cartesian product $A \times B$ can be shown as follows:

(a, g)　(b, g)　(c, g)
(a, f)　(b, f)　(c, f)
(a, e)　(b, e)　(c, e)
(a, d)　(b, d)　(c, d)

Each ordered pair of $A \times B$ is associated with a position in the lattice containing _____ columns and _____ rows.

3; 4

The Set of Whole Numbers

185. If $A = \{u, v, w\}$ and $B = \{r, s\}$, then $A \times B$ contains ___ elements. The associated lattice appears as

　　　　　o 　o 　o
　　　　　o 　o 　o

6

186. If $A = \{m, n, o\}$ and $B = \{k\}$, then $A \times B$ contains ___ elements. Draw the lattice.

3; o o o

187. If $A = \{k, l, m, n, o\}$ and $B = \{r\}$, then $A \times B$ contains ___ elements. Draw the lattice.

5; o o o o o

Remark. The previous two frames suggest that the product of any whole number and 1 is the identical (same) whole number. We assume that the following is always true:

　　　For all $a \in W$, $1 \cdot a = a$ and $a \cdot 1 = a$.

This axiom is called the **Identity law of multiplication.**

188. Thus, using this axiom, $7 \cdot 1 = 7$ and $1 \cdot 7 = $ ___ .

7

189. Recall that if an operation on two members in a set reproduces the second member, the first member is called an **identity element** for that operation. Since for all $a \in W$, $1 \cdot a = $ ___ , 1 is called the _____ element for multiplication in the set of whole numbers.

---

a; identity

Remark. Now we shall adopt additional axioms for the operation of multiplication similar to those we took for the operation of addition—the laws of closure, commutativity, and associativity.

190. Recall that when the result of operating on 2 elements of a given set is always another element in the same set, the set is said to be closed with respect to this
▼

operation. For example, 3 is a whole number, 7 is a whole number, and visualizing the appropriate Cartesian product or lattice, we would agree that the product, 3 · 7, is a _____ number.

---

whole

Remark. We shall assume that the product of any two whole numbers is a whole number and take the following axiom:

$$\text{For all } a,b \in W, \ a \cdot b \in W.$$

This statement is called the **Closure law for multiplication**.

191. Another way to express this law would be to assert that the set of whole numbers is _____ with respect to _____ .

---

closed; multiplication

Remark. Although the set of whole numbers is closed with respect to addition and multiplication, it is not closed with respect to other operations, subtraction and division, that we shall consider later in this program.

We have assumed that the addition of whole numbers is both commutative and associative. Do such assumptions seem warranted in the case of multiplication?

192. Is the product of 2 and 3 the same as the product of 3 and 2? Consider the sets $A = \{a, b\}$ and $B = \{x, y, z\}$.

$$A \times B = \{(a, x), (a, y), (a, z), (b, x), (b, y), (b, z)\}$$
and
$$B \times A = \{(x, a), (x, b), (y, a), (y, b), (z, a), (z, b)\}$$

contain the same number of elements. This example suggests that multiplication (is/is not) commutative.

---

is

Remark. Considering the general nature of Cartesian products similar to those in Frame 192, we make the following assumption:

$$\text{For all } a,b \in W, \ a \cdot b = b \cdot a.$$

This assertion is called the **Commutative law of multiplication**.

The Set of Whole Numbers

193. That (2417) · (7326) and (7326) · (2417) represent the same product is guaranteed by the Commutative law of multiplication. Since this law is an assumption, it can also be called an _____ .

axiom

194. Multiplication is a binary operation. To multiply the members of any given collection of numbers, the operation of multiplication is performed on only _____ numbers at a time.

two

195. Since multiplication is a binary operation, we want to give meaning to an expression such as $a \cdot b \cdot c$. We shall define $a \cdot b \cdot c$ to mean the product of $a \cdot b$ and c. That is,

$$\text{For all } a,b,c \in W, \quad a \cdot b \cdot c = (a \cdot b) \cdot c.$$

The operation of multiplication is first performed on the product _____ .

$a \cdot b$

Remark. We shall take the following assertion as an axiom.

$$\text{For all } a,b,c \in W, \quad (a \cdot b) \cdot c = a \cdot (b \cdot c).$$

This statement is called the **Associative law of multiplication**.

196. The fact that (2 · 3) · 4 and 2 · (3 · 4) are the same product follows from the _____ law of multiplication.

Associative

197. Besides the Identity law of multiplication, the three basic laws that apply to the multiplication of whole numbers are: the _____ law, the _____ law, and the _____ law.

Closure; Commutative; Associative *(In any order.)*

198. "For all $a,b \in W$, $a \cdot b \in W$." This statement is an expression of the _____ law for multiplication.

Closure

199. For all $a, b \in W$, $a \cdot b = b \cdot a$.

This is an expression of the _____ law of multiplication.

Commutative

200. For all $a, b, c \in W$, $(a \cdot b) \cdot c = a \cdot (b \cdot c)$.

This is an expression of the _____ law of multiplication.

Associative

201. Both the *Commutative law of addition* and the *Commutative law of multiplication* have to do with the *order* of numbers. Thus, when 4 + 6 is written 6 + 4, or when (4) · (6) is written (6) · (4), only the _____ of the numbers has been changed.

---
order

202. Both the *Associative law of addition* and the *Associative law of multiplication* have to do with *grouping*. Thus, when (4 + 8) + 3 is written 4 + (8 + 3), or when (4 · 8) · 3 is written 4 · (8 · 3), only the _____ of the numbers has been changed.

---
grouping

203. The law that permits writing (7) · (8) as (8) · (7) is the _____ ____ ____ _____ .

Commutative law of multiplication

204. The law that permits writing 8 + 6 as 6 + 8 is the _____ ____ ____ _____ .

Commutative law of addition

205. The law that permits writing (8 · 6) · 9 as 8 · (6 · 9) is the _____ ____ ____ _____ .

Associative law of multiplication

The Set of Whole Numbers

206. The law that permits writing (8 + 6) + 9 as 8 + (6 + 9) is the _____
_____ _____ _____.

Associative law of addition

Remark. So far we have been able to obtain a basic numeral for a product only by counting the elements in a Cartesian product or a lattice. Let us assume that we have now done this for all products whose factors are numbers from 1 to 10 and have summarized the facts in a table. As was the case earlier, when we introduced the table of addition facts, we can use the Commutative law to abbreviate the entries in the following table, since products such as 2 · 3 and 3 · 2 are identical.

·	1	2	3	4	5	6	7	8	9	10
1	1									
2	2	4								
3	3	6	9							
4	4	8	12	16						
5	5	10	15	20	25					
6	6	12	18	24	30	36				
7	7	14	21	28	35	42	49			
8	8	16	24	32	40	48	56	64		
9	9	18	27	36	45	54	63	72	81	
10	10	20	30	40	50	60	70	80	90	100

With this table available, we can replace any product, neither of whose factors is greater than 10, by a basic numeral which is an entry in the table.

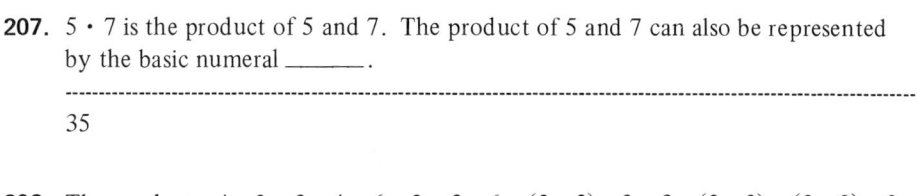

207. 5 · 7 is the product of 5 and 7. The product of 5 and 7 can also be represented by the basic numeral _____ .

35

208. The products 4 · 3, 3 · 4, 6 · 2, 2 · 6, (2 · 2) · 3, 3 · (2 · 2), (2 · 3) · 2, (3 · 2) · 2, 2 · (2 · 3), and 2 · (3 · 2) can all be represented by the _____ _____ 12.

---
basic numeral

Remark. A basic numeral may correspond to a number of different products, although any single product corresponds to just one basic numeral.

209. If $a = 3$, b is a whole number, and $a \cdot b = 12$, the Substitution law justifies writing $3 \cdot b = 12$. Similarly, if $r = 4$, s is a whole number, and $r \cdot s = 20$, then $4 \cdot \underline{} = 20$.

$4 \cdot s = 20$

210. If r and s are whole numbers, and if $r = s$ and $r \cdot t = 15$, then the $\underline{}$ law justifies writing $s \cdot t = 15$.

Substitution

Remark. Now, let us consider a consequence of our axioms for multiplication. Recall that we have the following theorem that followed from the axioms we had taken for the operation of addition.

$$\text{If } a, b, c \in W, \text{ and if } a = b, \text{ then } a + c = b + c.$$

A similar theorem can be stated for the operation of multiplication.

Theorem: If $a, b, c \in W$ and if $a = b$, then $a \cdot c = b \cdot c$.

Proof:

$a,b,c \in W$ and $a = b$	(given information)
$a \cdot c \in W$	(Closure law for multiplication)
$a \cdot c = a \cdot c$	(Reflexive law of equality)
$a \cdot c = b \cdot c$	(Substitution law; b substituted for a)

The last statement is exactly what we wished to prove.

211. The statement in the previous remark,

$$\text{If } a, b, c \in W \text{ and if } a = b, \text{ then } a \cdot c = b \cdot c,$$

is a theorem. Its truth follows from definitions we have made and the $\underline{}$ we have taken for the whole numbers.

---

axioms

212. The Commutative law of multiplication justifies writing $a \cdot c$ as $c \cdot a$ and $b \cdot c$ as $c \cdot b$. Therefore, $a \cdot c = b \cdot c$ can be written $c \cdot \underline{} = c \cdot \underline{}$.

$c \cdot a = c \cdot b$

The Set of Whole Numbers

213. If $a = b$, then $3 \cdot a = \underline{} \cdot b$.

$3 \cdot a = 3 \cdot b$

214. If $r = s$, then $5 \cdot r = \underline{} \cdot \underline{}$.

$5 \cdot r = 5 \cdot s$

Remark. We have seen that there are closure, commutative, associative, and identity laws for both addition and multiplication in the set of whole numbers. There is one more law that is fundamental to both operations and establishes a relationship between them. In order to introduce this law, it will be necessary to develop further the use of grouping devices in the process of multiplication.

215. Recall that parentheses, brackets, or other grouping symbols are used to group numbers if the group is to be considered as a single number. Thus, $(a + 4)$ means that the sum of a and 4 is to be thought of as a single _____ .

number

216. Thus, $3 \cdot (a + 2)$ or $(3) \cdot (a + 2)$ is the product of the numbers _____ and $(a + 2)$, where $a + 2$ is thought of as a single number.

3

217. $(4) \cdot (a + 3)$ is the _____ of the factors 4 and $(a + 3)$.

product

Remark. Note that through the use of the Commutative law for multiplication, the product $(4) \cdot (a + 3)$ can also be written as $(a + 3) \cdot (4)$.

218. The product of 3 and $(2a + b)$ is $3 \cdot (2a + b)$. Write the product of 7 and $(2a + b)$.

$7 \cdot (2a + b)$ $[(7) \cdot (2a + b)$ *is also correct. We shall use the two forms interchangeably.*]

219. Write the product of a and $(b + c)$.

$(a) \cdot (b + c)$

74 UNIT II

220. A basic numeral to represent the product of 3 · (1 + 4) can be obtained by first adding 1 and 4 to obtain 5 and then multiplying 3 by 5. The basic numeral is _____ .

15

221. 2 · (3 + 5) is the product of the numbers 2 and (3 + 5). An array illustrating this product can be constructed as follows:

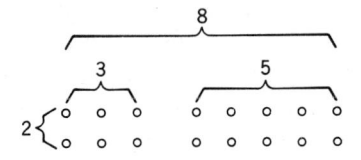

There are 16 elements in the array. The array can be considered to represent either the product 2 · (8) or the sum of the products 2 · (3) and 2 · (____).

5

Remark. The factor 2 in the previous frame is said to be *distributed* over each part of the sum 3 + 5.

222. In like manner, 5 · (1 + 4) = 5 · (____) + 5 · (____).

1; 4

223. 4 · (8 + 2) = ____ · (8) + ____ · (2).

4; 4

Remark. We can't check each possible relationship between sums and products of all whole numbers, therefore, we make the assumption:

For all a, b, c ∈ W, a · (b + c) = a · b + a · c.

This statement is called the **Distributive law.**

224. As a special case, 5 · (1 + 4) = 5 · (1) + 5 · (4) illustrates the _____ law.

Distributive

The Set of Whole Numbers 75

Remark. Observe the way in which this law is written. We have what might be termed a "left-hand" Distributive law because the factor a, which is distributed over the sum $(b + c)$, appears on the left. However, from the Commutative law of multiplication, we can write

$$a \cdot (b + c) = (b + c) \cdot a.$$

Hence, we also have a "right-hand" Distributive law. Therefore, if we want to write $3 \cdot a + 5 \cdot a = (3 + 5) \cdot a$, we will justify it by citing the Distributive law.

225. $3 \cdot (7 + 5) = 3 \cdot (7) + 3 \cdot (5)$ is an application of the left-hand _____ law.

Distributive

226. By the Distributive law, $2 \cdot (6 + 3) =$ ____ $\cdot (6) +$ ____ $\cdot (3)$.

2; 2

227. By the Distributive law, $(7 + 2) \cdot (4) = (7) \cdot$ ____ $+ (2) \cdot$ ____ .

4; 4

228. Based on the Distributive and Associative laws, it can be shown that $a \cdot (b + c + d + \cdots) = a \cdot b + a \cdot c + a \cdot d + \cdots$ for a finite number of terms in the parentheses. Thus,

$$4 \cdot (6 + 9 + 8) = 4 \cdot (\underline{}) + 4 \cdot (\underline{}) + 4 \cdot (\underline{}).$$

6; 9; 8

Remark. Recall that 0 is the only element in the set of whole numbers with the property that $a + 0 = a$. How does this property determine the property of 0 for multiplication?

229. Recall that if a is any whole number, the Identity law for _____ states:

$$a + 0 = a \text{ and } 0 + a = a.$$

---
addition

76   UNIT II

230. In particular, by the Substitution law, if $a = 0$, then, because $a + 0 = a$,

$$0 + 0 = \underline{}.$$

$0 + 0 = 0$

231. Now, if $0 + 0 = 0$, then the theorem if $a = b$, then $c \cdot a = c \cdot b$ assures us that

$$3 \cdot (0 + 0) = 3 \cdot 0,$$

and by the Distributive law, we can rewrite the left-hand member, $3 \cdot (0 + 0)$, and obtain

$$3 \cdot \underline{} + 3 \cdot \underline{} = 3 \cdot 0.$$

$3 \cdot 0 + 3 \cdot 0 = 3 \cdot 0$

232. The Identity law for addition says that there is exactly one number 0 that, when added to any integer, produces that integer for a sum. We have from the preceding frame that

$$3 \cdot 0 + \boxed{3 \cdot 0} = 3 \cdot 0,$$

and since $\boxed{3 \cdot 0}$ is a number such that, when added to $3 \cdot 0$, the result is $3 \cdot 0$, it must be the identity element for addition, that is 0. Therefore,

$$3 \cdot 0 = \underline{}.$$

$3 \cdot 0 = 0$

Remark. The previous frames suggest the theorem:

Theorem: If $a \in W$, then $a \cdot 0 = 0$.

The proof follows the same arguments as in Frames 229-232 above.

Proof:

$0 + 0 = 0$ (Identity law for addition)
$a \cdot (0 + 0) = a \cdot 0$ (Theorem: if $a = b$, then $c \cdot a = c \cdot b$)
$a \cdot 0 + \boxed{a \cdot 0} = a \cdot 0$ (Distributive law)
$\boxed{a \cdot 0} = 0$ (Identity law for addition)

233. Since $a \cdot 0 = 0$ for every whole number a, then $0 \cdot a = 0$, by the (Closure/Commutative/Associative) law of multiplication.

Commutative

The Set of Whole Numbers 77

234. 9 · 0 = ___ and 0 · 9 = ___ .

0; 0

235. In general, for all $a \in W$, $a \cdot 0 = $ ___ and $0 \cdot a = $ ___ .

$a \cdot 0 = 0$; $0 \cdot a = 0$

Remark. The Distributive law, which we take as an axiom, is the foundation upon which many of the algorithms of arithmetic rest. In the following sequence of frames, we shall try to give you an idea of why the Distributive law is the basis for certain familiar steps in the process of multiplication.

236. Note that 8 · (17) can also be written as 8 · (10 + 7). By the _____ law, this can be shown as 8 · (10) + 8 · (7), or 80 + 56, or the basic numeral ___ .

Distributive; 136

237. 9 · (23) can be written as 9 · (20 + 3). Write this product as the sum of two products.

---
9 · (20) + 9 · (3)   (*Or* 180 + 27.)

Remark. We have (page 71) a 10 by 10 multiplication table, but it does not contain a value for 9 · 20. To avoid getting involved in a discussion of numeration, however, we shall assume (as we did with our addition table earlier) that if you know the product of 9 and 2, then you know the product of 9 and 20, 9 and 200, 90 and 2, and so on. What we are concerned with here is the application of the Distributive law in the algorithm.

238. 4 · (764) can be written as 4 · (700 + 60 + 4). Write this product as the sum of three products.

---
4 · (700) + 4 · (60) + 4 · (4)   (*Or* 2800 + 240 + 16.)

239. Using the vertical form for multiplication,

```
   238
  × 9
```

▼

can be written

$$(200 + 30 + 8)$$
$$\times \qquad 9$$

Write this product as the sum of three products.

$9 \cdot (200) + 9 \cdot (30) + 9 \cdot (8)$ (Or $1800 + 270 + 72$.)

Remark. In multiplying 238 by 9, we used $9 \cdot (200) + 9 \cdot (30) + 9 \cdot (8)$ or $1800 + 270 + 72$. To help us discuss such expressions, we shall call each of the terms $9 \cdot (200), 9 \cdot (30)$, and $9 \cdot (8)$ [or 1800, 270, and 72] *partial products.*

240. The product $7 \cdot (6032)$ can be changed to $7 \cdot (6000 + 30 + 2)$. When this is written as the sum of three partial products, $7 \cdot (6000) + 7 \cdot (30) + 7 \cdot (2)$, the _____ law has been applied.

Distributive

241. The product $(14) \cdot (12)$ can be written as $(10 + 4) \cdot (12)$. Application of the Distributive law yields $10 \cdot (\quad) + 4 \cdot (\quad)$.

$10 \cdot (12) + 4 \cdot (12)$

242. Going one step further, $10 \cdot (12) + 4 \cdot (12)$ can be written as $10 \cdot (10 + 2) + 4 \cdot (10 + 2)$. An application of the Distributive law to each of these two partial products gives the sum of four partial products,

_____ + _____ + _____ + _____ .

$10 \cdot (10) + 10 \cdot (2) + 4 \cdot (10) + 4 \cdot (2)$

243. If we compare this result with the product of 14 and 12 shown in detail in vertical form, we can readily see the similarity.

$$\begin{array}{r} 12 \\ \times\ 14 \\ \hline 48 = 4 \cdot (10) + 4 \cdot (2) \\ 120 = 10 \cdot (10) + 10 \cdot (2) \end{array}$$

The sum of the four partial products can be written as the basic numeral _____ .

168

The Set of Whole Numbers 79

Remark. In many areas of mathematics, a horizontal rather than a vertical representation of sums and products is easier to employ and much preferred. In general, we will use this style in examples that follow.

244. The product (17) · (21) can be written (10 + 7) · (20 + 1). One application of the Distributive law, thinking of (10 + 7) as a single number, yields (10 + 7) · (20) + (10 + 7) · (1). A second application of the Distributive law to each of these partial products gives the sum of four partial products, ____ + ____ + ____ + ____ .

---
10 · (20) + 7 · (20) + 10 · (1) + 7 · (1)

245. The product (14) · (19) can be written (10 + 4) · (10 + 9). An application of the Distributive law yields (10 + 4) · (10) + (10 + 4) · (___).

9

246. Apply the Distributive law to write (10 + 4) · (10) + (10 + 4) · (9) as the sum of four partial products.

---
(10) · (10) + (4) · (10) + (10) · (9) + (4) · (9)

247. The product (23) · (16) can be written (20 + 3) · (10 + 6). Write this product as the sum of four partial products by successive applications of the Distributive law.

---
(20) · (10) + (3) · (10) + (20) · (6) + (3) · (6)

Remark. We have now considered some basic properties for whole numbers. These properties establish the foundation for our work with the set of whole numbers and more extensive sets in the work that follows.

The panel on page 81 summarizes the axioms that we have adopted and some consequences that followed from these axioms. You may want to refer to these statements as you proceed through the remainder of the program.

Panel I

$a, b, c \in W$

AXIOMS OF EQUALITY

1. $a = a$.   Reflexive law of equality
2. If $a = b$, then $b = a$.   Symmetric law of equality
3. If $a = b$ and $b = c$, then $a = c$.   Transitive law of equality
4. If $a = b$, then a may be substituted for b and b for a in any expression without changing the truth or falsity of the statement.   Substitution law of equality

AXIOMS OF ORDER

1. Exactly one of the following relations holds: Trichotomy law
   $a < b$, $a = b$, or $b < a$.
2. If $a < b$ and $b < c$, then $a < c$.   Transitive law of inequality

AXIOMS FOR OPERATIONS

$a + b \in W$.   Closure law for addition
$a + b = b + a$.   Commutative law of addition
$(a + b) + c = a + (b + c)$.   Associative law of addition
$a + 0 = a$ and $0 + a = a$.   Identity law of addition
$a \cdot b \in W$.   Closure law for multiplication
$a \cdot b = b \cdot a$.   Commutative law of multiplication
$(a \cdot b) \cdot c = a \cdot (b \cdot c)$.   Associative law of multiplication
$a \cdot 1 = a$ and $1 \cdot a = a$.   Identity law of multiplication
$a \cdot (b + c) = a \cdot b + a \cdot c$.   Distributive law

PROPERTIES

If $a = b$, then $a + c = b + c$.
If $a = b$, then $a \cdot c = b \cdot c$.
$a \cdot 0 = 0$.

The Set of Whole Numbers   83

Remark. The following frames will give you an opportunity to review some of the fundamental properties related to the operation of multiplication on the whole numbers.

R248. The operation of multiplication on two whole numbers, pairs the numbers, called factors, with a third unique number, called the _____ of the numbers.
----
product

R249. $A \times B$ is called the _____ product of A and B.

Cartesian

R250. The product $r \cdot s$ can be viewed as the number of elements in a _____ formed by taking r columns and s rows.
----
lattice (*Or array.*)

R251. If an operation on two members in a set reproduces the second member, the first member is called the _____ element for that operation.
----
identity

R252. The number 1 has the following property;
$$\text{For all } a \in W, \ 1 \cdot a = a \ \text{ and } \ a \cdot 1 = a.$$
The number 1 is called the _____ element for _____.
----
identity; multiplication

R253. The assumption that the product of two whole numbers is always another whole number is called the _____ law for multiplication.

Closure

R254. The (Closure/Commutative/Associative) law of multiplication justifies writing $384 \cdot 15$ as $15 \cdot 384$.

Commutative

R255. The (Closure/Commutative/Associative) law of multiplication justifies writing $(3 \cdot 8) \cdot 5$ as $3 \cdot (8 \cdot 5)$.

Associative

R256. The _____ law of multiplication justifies writing $7 \cdot 1 = 7$ and $1 \cdot 7 = 7$.

Identity

R257. Complete the following theorem:

If $a, b, c \in W$, and if $a = b$, then $a \cdot c =$ _____.

$a \cdot c = b \cdot c$

R258. Writing a product $7 \cdot (10 + 3)$ as the sum $7 \cdot (10) + 7 \cdot (3)$ is an application of the _____ law.

Distributive

R259. For all $a \in W$, $a \cdot 0 =$ ____ and $0 \cdot a =$ ____.

$a \cdot 0 = 0;\ 0 \cdot a = 0$

Remark. See Exercise IId, page 252, for additional practice on the operation of multiplication.

We are now in a position to discuss what is meant by the term **number system** in the title of this program.

A number system consists of:
1. A set of elements;
2. At least one well-defined operation on the elements of the set;
3. Some assumptions (axioms) that govern the operation or operations; and
4. Logical consequences of the axioms.

Now, the number system we have been discussing is the system of whole numbers. We have used $W = \{0, 1, 2, 3, \ldots\}$, the operations of addition and multiplication, and some equality and order properties, together with some assumptions about the results of the operations to see what these things implied. Our treatment has been far from exhaustive.

There are operations that *are not* necessarily *basic* to a number system, but they enhance its usefulness. The first of these operations that we shall look at briefly is **subtraction**.

The Set of Whole Numbers   85

260. Consider the sum 5 + 4 = 9, in which 4 is the number that when added to 5 equals 9. The number 4 is called the **difference** of 9 and 5. Similarly, 2 is the difference of 5 and 3 because 2 added to ____ equals ____.

3; 5

261. The number 7 is the _____ of 10 and 3 because the sum of 7 and 3 is 10.

---
difference

262. Consider 10 − 3, the difference of 10 and 3. We can write 10 − 3 as 7, the number that added to 3 equals 10. We can write 8 − 5, the _____ of 8 and 5, as the basic numeral 3.

---
difference

Remark. We define a difference as follows:

For all a,b ∈ W, a − b is the whole number d (if one exists) such that when d is added to b, the sum equals a.

263. From the definition of a difference, $a - b = d$ if $\underline{b + d =}$ ____ .

$b + d = a$

264. Consider $3 + d = 5$, where d is the number that _____ to 3 the sum equals 5. The number d is called the _____ of 3 subtracted from 5 and can be represented by 5 − 3 or the basic numeral 2.

---
added; difference

265. The _____ of 12 and 9 can be written as 12 − 9 or the basic numeral 3.

---
difference

Remark. The difference 12 − 9 is the whole number that when added to 9, the sum equals 12. In the previous frame we represented this number by the basic numeral 3. But what is the meaning of 9 − 12? Does this expression name a number in the set of

▼

86   UNIT II

whole numbers? Is there a whole number that when added to 12 yields 9? Obviously not, but let us explore this question further by using the number line.

266. Natural numbers can be associated with the *counting* of marked points *to the right* of the origin on the number line. The number line

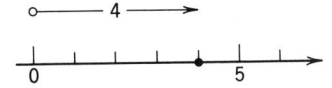

illustrates how 4 can be associated with _____ four marked points to the _____ of the origin.

---

counting; right

267. The sum 4 + 2 can be thought of as the number associated with the point reached after counting off four of the chosen segments to the right of the origin, and then, *from this point*, counting off two more marked points to the right.

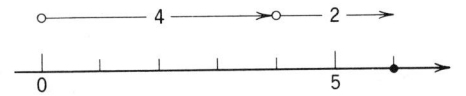

The number associated with the point arrived at last is _____ .

---

6 (*Or 4 + 2, since these symbols also represent the number "six."*)

268. Show the addition of 5 and 3 on the given number line.

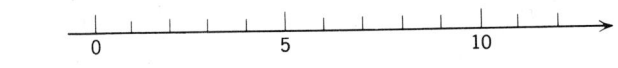

---

269. Subtraction of natural numbers can be associated with *counting to the left* on a number line. Thus, 5 − 3 is the number associated with the position arrived at after counting five segments to the right, and then, from this point, counting three segments to the left, as follows:

▼

The Set of Whole Numbers 87

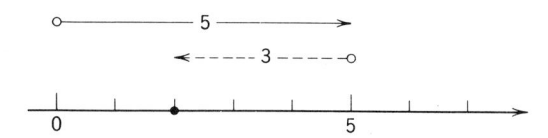

The number associated with the point arrived at last is ____.

---

2 (Or 5 − 3.)

270. Use the given number line to show how the difference 6 − 2 can be arrived at.

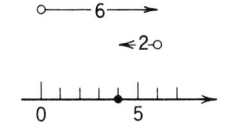

---

271. Now, the big question. How would a difference such as 9 − 12 appear on the number line? Using our counting procedure, it would appear as

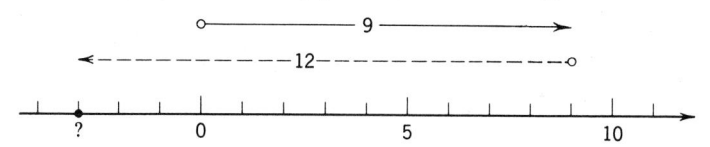

Since *the whole numbers are associated only with points at the origin or to the right of the origin*, 0, the difference 9 − 12 (is/is not) a member of the set of whole numbers.

---

is not

272. Again, if a and b are whole numbers, and if $a > b$, the difference $a - b$ would appear on a number line as

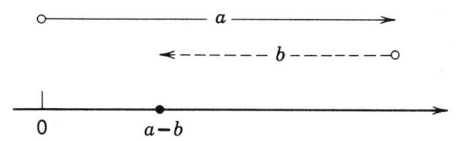

The difference $a - b$ (is/is not) a whole number.

---

is

273. If $a < b$, the difference $a - b$ would appear on the left extension of the number line as

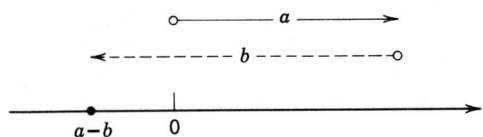

```
         o————— a —————>
   <———————— b ————————o
   ————————•——|————————————>
           a-b  0
```

The difference $a - b$ (is/is not) a whole number and shows that the set of whole numbers (is/is not) closed with respect to the operation of subtraction.

-----

is not; is not

274. If our universe consists only of whole numbers, $a - b$, where $a < b$, has no meaning for us, since it does not belong in our universe. Thus, we say that $6 - 19$, for example, has no meaning or is **undefined** in the set of whole numbers. Similarly, $8 - 25$ is _____ in the set of whole numbers.

-----

undefined

Remark. We have seen that *the set of whole numbers is not closed with respect to subtraction.* $a - b$ is a whole number only if $a \geq b$. We shall be able, however, to give meaning to the symbol $a - b$ for every pair of whole numbers when we enlarge our universe later in the program.

In the meantime, this limitation does not prevent us from subtracting one whole number from another when this is possible.

We are now ready to examine the last of our four operations—that of **division**. Recall for a moment that the process of multiplication can be viewed as a process of repeated addition. The operation of division can be viewed as a process of repeated subtraction. To ask someone to divide 21 by 7 can be interpreted as asking how many times 7 can be "taken out of," or subtracted from, the number 21. We intend, however, to approach division and the meaning of a quotient over a different route in order to lay the groundwork for the introduction of a new set of numbers later in the program.

275. Consider the equality $5 \cdot 4 = 20$. The number 20 is the product of 5 and 4. 4 is said to be the **quotient** of 20 divided by 5. Similarly, $2 \cdot 3 = 6$ and 3 is the _____ of 6 _____ by 2.

-----

quotient; divided

The Set of Whole Numbers

Remark. We shall define a quotient as follows:

For all $a, b \in W$, the quotient a divided by b is the whole number q (if one exists), such that $b \cdot q = a$.

276. For example, the quotient of 15 divided by 5 is the whole number q such that $5 \cdot q = 15$. In this example, q represents the basic numeral ____.

3

277. The quotient of a divided by b can be written as $a \div b$ or as the fraction $\frac{a}{b}$. Thus, $6 \div 2$ can be written in fraction form as ____.

$\frac{6}{2}$

278. The fraction $\frac{10}{2}$ represents the quotient of 10 divided by 2. The quotient is equal to 5 because 2 times ____ equals 10.

5

279. The fraction $\frac{12}{4}$ represents the ____ of 12 divided by 4.

---
quotient

280. The fraction $\frac{12}{4}$ equals ____ because 4 times ____ equals 12.

3; 3

281. From our definition of a quotient, for all $a, b \in W$, $\frac{a}{b}$ is the whole number, q, (if one exists) such that $b \cdot$ ____ $= a$.

$b \cdot q = a$

282. By the definition of division, if a is a whole number, $\frac{a}{1}$ equals ____ because $1 \cdot$ ____ $= a$.

---
a; a

UNIT II

283. $\frac{a}{1}$ and a represent the same number; $\frac{3}{1}$ and 3 represent the _____ number.

same

284. $\frac{12}{2}$ and 6 are symbols that represent the _____ number.

same

Remark. We have seen that various symbols may represent the same number. For example, "4 + 2", "3 · 2", "8 – 2", $\frac{"12"}{2}$, and "6" are symbols that represent the same number. The numeral "6" is called the **basic numeral** and, in general, is the symbol we prefer to use.

Up to this point, all of the division problems we have examined have resulted in quotients that are in the set of whole numbers. This is not always the case.

285. The symbol $\frac{5}{3}$ is meaningless in the system of whole numbers because there is no whole number whose product with 3 equals 5. The symbol $\frac{8}{4}$ (does/does not) denote a whole number. The symbol $\frac{8}{3}$ (does/does not) denote a whole number.

does; does not

286. Recall that when a symbol is meaningless in a number system, it is said to be undefined in the system. The symbol $\frac{9}{5}$ is undefined in the system of whole numbers because it does not represent a _____ number.

whole

287. The symbol $\frac{9}{3}$ is defined in the system of whole numbers. The symbol $\frac{9}{5}$ is _____.

undefined   (Or *meaningless.*)

288. According to the definition of a quotient, $\frac{3}{0}$ would have to be a whole number q such that $0 \cdot q = 3$. But we have seen that for each whole number q, $0 \cdot q = 0$.

▼

The Set of Whole Numbers   91

Therefore, $\frac{3}{0}$ (does/does not) represent a whole number.

---

does not

289. In short, $\frac{3}{0}$ does not represent a member of the set of whole numbers, because there is no whole number q such that $0 \cdot q = 3$. Similarly, $\frac{5}{0}$ (does/does not) represent a member of the set of whole numbers.

---

does not

290. Now, consider the symbol $\frac{0}{0}$. The definition of a quotient requires that $\frac{0}{0}$ represent the whole number q such that $0 \cdot q = 0$. Notice that q may be any whole number. That is, $\frac{0}{0}$ could be 1, or 2, or etc. This ambiguous situation means, therefore, that $\frac{0}{0}$ can be interpreted to be equal to any _____ .

---

whole number   (*Or number.*)

291. $\frac{0}{0}$ is an ambiguous symbol because it can represent any element in the set of whole numbers. For any whole number b, not equal to 0, the symbol $\frac{b}{0}$ is meaningless because it does not represent a member in the set of _____ .

---

whole numbers

292. It appears from the preceding frames that for the elements in the set of whole numbers we must restrict our division operation to exclude division by _____ .

---

zero

293. Since division by zero either leads to an ambiguous situation or fails to yield a number in our universe, division by zero is said to be undefined. Thus, the number zero has the properties that $3 + 0 = 3$, $3 \cdot 0 = $ ____ , and $\frac{3}{0}$ is _____ .

0; undefined   (*Or meaningless*)

294. On the other hand, the quotient of 0 divided by 3, that is, $\frac{0}{3}$, is the whole number 0 because 3 · 0 = 0. While $\frac{3}{0}$ is _____, $\frac{0}{3}$ is equal to 0.

---
undefined

295. If a and b are whole numbers, $\frac{a}{b}$ is not always a member of the set of whole numbers. Therefore, the set of whole numbers (is/is not) closed with respect to division.

---
is not

Remark. We see now that the set of whole numbers is not closed with respect to subtraction and division. Later in this program, however, you will see how it will be possible to give more meaning to the symbols $a - b$ and $\frac{a}{b}$ ($b \neq 0$) for every whole number a and b, if numbers other than whole numbers are included in our universe.

The following sequence of frames will give you an opportunity to review ideas concerning the operations of subtraction and division.

R296. The difference $a - b$, for all $a,b \in W$ and $a \geqslant b$, is the whole number d such that ____ + d = ____ .

$b + d = a$

R297. For all $a,b \in W$, the difference $a - b$ is a whole number if a ____ b and is not a whole number if a ____ b.

$a \geqslant b$; $a < b$

R298. Show how the difference $7 - 4$ can be found by using the line

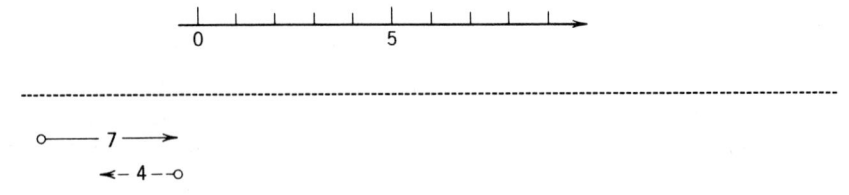

The Set of Whole Numbers 93

R299. The set of whole numbers (is/is not) closed with respect to addition and (is/is not) closed with respect to subtraction.

---
is; is not

R300. The quotient of a divided by b where a and b are whole numbers is a number q (if one exists) such that ____ $\cdot\, q =$ ____ .

$b \cdot q = a$

R301. The symbol $\frac{a}{b}$ is the indicated quotient of a and b and is called a ____.

---
fraction

R302. $18 \div 3$ can be written in fraction form as $\frac{18}{3}$. The quotient $\frac{18}{3}$ is a number (which can be written as the basic numeral ____) such that $3 \cdot$ ____ $= 18$.

6; $3 \cdot 6 = 18$

R303. If a is a whole number, $\frac{a}{1}$ equals ____ because $1 \cdot$ ____ $= a$.

---
a; $1 \cdot a = a$

R304. The symbol $\frac{6}{2}$ (does/does not) denote a whole number; the symbol $\frac{6}{5}$ (does/does not) denote a whole number.

---
does; does not

R305. The symbol $\frac{5}{0}$ (does/does not) represent a whole number.

---
does not

R306. The symbol $\frac{0}{5}$ represents the number ____ .

0

R307. The set of whole numbers (is/is not) closed with respect to multiplication and (is/is not) closed with respect to division.

is; is not

Remark. See Exercise IIe, page 255, for additional practice on the operations of subtraction and division.

We have come to the end of our discussion of the system of whole numbers. In addition to the basic operations of addition and multiplication in the system, we have now defined in terms of these, two operations, subtraction and division. The inclusion of these two operations renders our system much more useful as a model for real-life situations, even though the set of whole numbers is not closed for these operations.

The following sequence of frames will provide you an opportunity to review all the basic ideas in Unit II of this program, The Set of Whole Numbers. You may want to refer to Panel I on page 81 as you complete the review.

R308. The property shared by all sets equivalent to a given set when all properties of the sets are completely disregarded is called _____.

number

R309. When a whole number is used to describe the number of members in a set, it is being used in a cardinal sense. When such a number is used to describe the position of a member in a set, it is being used in an _____ sense.

ordinal

R310. Symbols such as 2, 3, and 7, which represent specified elements in the set of whole numbers, are called numerals. Symbols such as *a, b, c, x, y, z*, which refer to any member of a set without specifying a particular member of the given set, are called _____.

variables

R311. All variables in this part of the program have represented _____ numbers.

whole

R312. A variable represents any element of a given set of numbers. The set of numbers is called the _____ set of the variable.

replacement   (*Or universal*)

The Set of Whole Numbers 95

R313. $\{a \mid a > 10, a \in W\}$ illustrates the use of set-builder notation. The vertical line is read "_____ _____."

such that

R314. Whole numbers can be associated with points on a line. The point is called the _____ of the number and the number is called the _____ of the point.

graph; coordinate

R315. On the number line

7 is the _____ of the point labeled A. 0, 5 and 10 are scale numbers.

coordinate

Remark. Recall that we made four agreements concerning the use of symbolism pertaining to equality. These are: the Reflexive law, the Symmetric law, the Transitive law, and the Substitution law.

R316. We agreed that a symbol will always represent the same number during the course of a particular discussion. That is:

$$\text{For all } a \in W, \quad a = \underline{\quad}.$$

This statement is called the _____ law of equality.

$a = a$; Reflexive

R317. We agreed that if two different symbols are used for the same number, then the order in which a statement of equality involving these symbols is written doesn't matter. That is:

$$\text{For all } a, b \in W, \text{ if } a = b, \text{ then } \underline{b} = \underline{\quad}.$$

This statement is called the _____ law of equality.

$b = a$; Symmetric

R318. We agreed that if the symbols *a* and *b* denote the same number and if the symbols *b* and *c* denote the same number, then the symbols *a* and *c* denote the same number. That is:

For all $a, b, c \in W$, if $a = b$ and $b = c$, then $\underline{a =}$.

This statement is called the _____ law of equality.

$a = c$; Transitive

R319. We agreed that:

If two symbols denote the same number, one of the symbols may be substituted for the other in any expression without changing the truth or falsity of the statement.

This statement is an expression of the _____ law.

Substitution

R320. Writing $2 = a$ as $a = 2$ is justified by the _____ _____ of equality.

Symmetric law

R321. If $r = s$ and $s = 3$, then by the _____ _____ of equality, we may write $r = 3$.

Transitive law (*Or Substitution law*)

R322. For all $a, b, c \in W$, match each statement with the appropriate property.

1. If $b = c$, then $c = b$ a. Reflexive law of equality
2. $b = b$ b. Symmetric law of equality
3. If $c = b$ and $b = a$, then $c = a$ c. Transitive law of equality

---

1. - b.; 2. - a.; 3. - c.

R323. Formally stated assumptions such as the Reflexive, Symmetric, Transitive, and Substitution laws are called _____ .

---

axioms

The Set of Whole Numbers

Remark. Recall that we made two agreements concerning the order of whole numbers. These are: the Trichotomy law and the Transitive law.

R324. We assumed that for all whole numbers a and b, exactly one of the following is true: $a < b$, $a = b$, or _____. This statement is called the _____ law.

$b < a$; Trichotomy

R325. We assumed that if one whole number is less than a second and the second is less than a third, then the first is less than the third. That is:

For all $a, b, c \in W$, if $a < b$ and $b < c$, then ____ $<$ ____.

This statement is called the _____ law of inequality.

$a < c$; Transitive

R326. If $a < 3$ and $3 < x$, then by the _____ _____ of inequality we may write $a < x$.

Transitive law

R327. If r and s represent whole numbers, then by the _____ _____ we can assert that either $r < s$, $r = s$, or $s < r$.

Trichotomy law

R328. Since the Trichotomy law and Transitive law of inequality are assumptions, they are postulates or _____.

---

axioms

R329. $A = \{h, i, j\}$ and $B = \{k, l\}$ are disjoint sets because they contain no common elements. The union of A and B, $A \cup B = \{h, i, j, k, l\}$. The number associated with A is 3, the number associated with B is ____, and the number associated with $A \cup B$ is ____.

2; 5

UNIT II

R330. The operation of addition is a binary operation and pairs two numbers with a third unique number. The number associated with the union of two disjoint sets is called the _____ of the numbers associated with each set.

---
sum

Remark. Recall that we took four axioms for the operation of addition for whole numbers. They are the Closure law, the Commutative law, the Associative law, and the Identity law.

R331. The sum of any two whole numbers is assumed to be a whole number. That is:

For all $a, b \in W$, $a + b \in W$.

This statement is called the _____ law for addition.

Closure

R332. The sum of any two whole numbers is assumed to be the same regardless of the order in which the numbers are added. That is:

For all $a, b \in W$, $a + b =$ ___ $+$ ___ .

This statement is called the _____ law of addition.

$a + b = b + a$; Commutative

R333. Because addition is a binary operation, we defined a collection of symbols such as $a + b + c$ to be equal to $(a + b) + c$. Furthermore, it was assumed that:

For all $a, b, c \in W$, $(a + b) + c = a + ($ ___ $+$ ___ $)$.

This statement, which is concerned with the grouping of terms in a sum, is called the _____ law of addition.

$a + (b + c)$; Associative

R334. The law that permits writing $(2 + 3) + 4$ as $2 + (3 + 4)$ is the _____ _____ of addition.

Associative law

The Set of Whole Numbers 99

R335. The law that permits writing 6 + 9 as 9 + 6 is the _____ _____ of addition.

Commutative law

R336. The law that guarantees that $a + 9$ is a whole number for every whole number replacement for a is called the _____ _____ for addition.

Closure law

R337. If a and b are whole numbers, and if $a = b$ and $a + 2 = 9$, then the _____ law justifies writing $b + 2 = 9$.

Substitution

R338. From the _____ law of addition, $5 + 0 = 5$, and $0 + 17 = 17$.

Identity

R339. Assumptions we make concerning numbers and operations on numbers are called axioms. Logical consequences of these axioms are called _____ .

---

theorems

R340. One consequence of our axioms asserts that if a, b, and c are whole numbers, and if $a = b$, then $a + c = b + c$. Thus, if $a = b$, then $a + 2 =$ ___ $+$ ___ .

$b + 2$ (Or $2 + b$.)

R341. The operation of multiplication pairs two whole numbers, called factors, with a third unique whole number called the _____ of the factors.

---

product

R342. The operation of multiplication on whole numbers can be visualized through the use of sets. Recall that if $A = \{k, l, m\}$ and $B = \{o, p\}$, then
▼

100   UNIT II

$A \times B = \{(k, o), (k, p), (l, o), (l, p), (m, o), (m, p)\}$. Each member of A is paired with each member of B. $A \times B$ is called the _____ product of A and B.

Cartesian

R343. The Cartesian product $R \times S = \{(a, b), (a, c)\}$ contains 2 members. A Cartesian product is a set and each member of the set is an ordered pair. The braces $\{\ \}$ denote a set. The parentheses, and comma (,), denote an _____ _____.

---

ordered pair

R344. If $M = \{x, y\}$ and $N = \{s, t\}$, then $M \times N = \{$ _____ $\}$. $M \times N$ contains _____ elements.

$\{(x, s), (x, t), (y, s), (y, t)\}$; 4

R345. If $A = \{a, b\}$, $B = \{r, s, t\}$, and $C = \{w, x, y, z\}$, then $A \times B$ contains _____ ordered pairs, $A \times C$ contains _____ ordered pairs, and $B \times C$ contains _____ ordered pairs.

6; 8; 12

R346. The product of the two whole numbers associated with sets A and B equals the number associated with _____ × _____, the _____ product of A and B.

$A \times B$; Cartesian

R347. The graph of $A \times B$ in Frame 345 can be represented by a _____ or an array, such as:

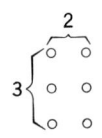

$$3\begin{cases} \overbrace{\begin{matrix} o & o \\ o & o \\ o & o \end{matrix}}^{2} \end{cases}$$

▼

The Set of Whole Numbers 101

This array contains ____ elements.

---
lattice; 6

R348. The set of whole numbers contains an element 1 having the following property.

$$\text{For all } a \in W, \ 1 \cdot a = \underline{\quad} \text{ and } a \cdot 1 = \underline{\quad}.$$

This statement is called the Identity law of _____.

---
a; a; multiplication

R349. 1 is called the _____ element for the operation of multiplication in the set of whole numbers.

---
identity

R350. If a is a whole number, the _____ for multiplication asserts that $a \cdot 7$ is a whole number.

Closure law

R351. If $7 \cdot 2$ is written as $2 \cdot 7$, the order of the factors has been changed. The _____ of multiplication asserts that these products are the same.

Commutative law

R352. Multiplication is a binary operation. If $(7 \cdot 3) \cdot 5$ is written as $7 \cdot (3 \cdot 5)$, the grouping of the factors has been changed. The _____ of multiplication asserts that these products are equal.

Associative law

R353. From the _____ law of multiplication, $8 \cdot 1 = 8$ and $1 \cdot 19 = 19$.

Identity

R354. The fact that the product $5 \cdot (10 + 2)$ can be written as the sum of the partial products $5 \cdot 10 + 5 \cdot 2$ is justified by the _____ law.

Distributive

R355. For all whole numbers a, b, and c, the Distributive law asserts that the product $a \cdot (b + c)$ can be written ____ $\cdot$ ____ $+$ ____ $\cdot$ ____ .

$a \cdot b + a \cdot c$

R356. A consequence of our axioms asserts that if a, b, and c are whole numbers and if $a = b$, then $a \cdot c = b \cdot c$. Thus, if $a = b$, then $a \cdot 2 =$ ____ $\cdot$ ____ .

$a \cdot 2 = b \cdot 2$ (Or $2 \cdot b$.)

R357. If $a \in W$, then $a \cdot 0 =$ ____ and $0 \cdot a =$ ____ .

$a \cdot 0 = 0$; $0 \cdot a = 0$

R358. Recall that addition is a basic operation in the system of whole numbers. Subtraction can be defined in terms of addition. Thus, $7 - 3$ is a number that when added to 3, the sum equals ____.

7

R359. The difference $7 - 3$ is a whole number. The symbol $3 - 7$ does not represent a whole number and is said to be _____ in the system of whole numbers.

---

undefined   (Or meaningless.)

R360. If a and b are whole numbers, $a - b$ is a whole number if a is (less/greater) than b. If a is (less/greater) than b, then $a - b$ is undefined.

---

greater; less

R361. The difference of two whole numbers is (always/sometimes/never) a whole number.

---

sometimes

R362. Multiplication is a basic operation in the system of whole numbers. Division can be defined in terms of multiplication. Thus $\dfrac{12}{4}$ is a number that,

▼

The Set of Whole Numbers   103

when multiplied by 4, the product equals _____. $\frac{25}{5}$ is a number that, when multiplied by 5, the product equals _____.

12; 25

R363. The fraction $\frac{12}{4}$ represents a whole number. The fraction $\frac{4}{12}$ does not represent a whole number and is said to be _____ in the system of whole numbers.

---

undefined   (*Or meaningless.*)

R364. If *a* and *b* are whole numbers, ($b \neq 0$) $\frac{a}{b}$ is a whole number *q* if *bq = a*. Thus $\frac{21}{8}$ (does/does not) represent a whole number and $\frac{21}{7}$ (does/does not) represent a whole number.

---

does not; does

R365. The set of whole numbers (is/is not) closed with respect to the operation of division.

---

is not

Remark. A self-evaluation test on Unit II of this program follows.

UNIT II

Self-Evaluation Test

Score your paper from the answers given on page 278.

1. The set of whole numbers is a(n) (finite/infinite) set.
2. $a < b$ means that the whole number denoted by a is _____ _____ the number denoted by b.
3. A number used to refer to the position of a member in an ordered set is said to be used in an _____ sense.
4. A symbol such as a, b, or c, used to represent an unspecified element in a specified set, is called a _____.
5. The number corresponding to a point on the number line is called the _____ of the point.
6. The (Reflexive/Symmetric/Transitive) law is an agreement that a given variable will not be used to represent different numbers during the course of a discussion.
7. "For all $a, b \in W$, if $a = b$, then $b = a$" is a statement of the (Reflexive/Symmetric/Transitive) law of equality.
8. "For all $a, b, c \in W$, if $a = b$ and $b = c$, then $a = c$" is a statement of the (Reflexive/Symmetric/Transitive) law of equality.
9. In mathematics, a formally stated assumption is called a postulate. Such an assumption is also called a _____.
10. "For all $a, b, c \in W$, if $a < b$ and $b < c$, then $a < c$" is called the (Trichotomy/Transitive) law of inequality.
11. If A and B represent disjoint sets, then the number associated with $A \cup B$ is called the _____ of the numbers associated with A and B.
12. When the result of adding two elements of a given set is always a unique element in the same set, the set is said to be _____ with respect to addition.
13. The _____ law of addition justifies writing $(2 + 4) + 6$ as $2 + (4 + 6)$.
14. The _____ law of addition justifies writing $7 + 16$ as $16 + 7$.

The Set of Whole Numbers 105

15. The _____ law of addition asserts that for all whole numbers a, $a + 0 = a$ and $0 + a = a$.

16. In mathematics, statements of logical consequences of definitions and axioms are called _____ .

17. If $a, b, c \in W$, and $a = b$, then $a + c$ (always/sometimes/never) equals $b + c$.

18. If $A = \{a\}$ and $B = \{b,c\}$, then $\{(a,b), (a,c)\} = (A \cup B \ / \ A \cap B \ / \ A \times B)$.

19. If $A = \{a, b, c\}$ and $B = \{d, e, f, g, h\}$, then $A \times B$, the Cartesian product of A and B, contains _____ ordered pairs.

20. The graph of a Cartesian product is called an array or a _____ .

21. The number 1 is the _____ element for multiplication.

22. "For all $a,b \in W$, $a \cdot b \in W$" is a statement of the _____ law for multiplication.

23. "For all $a,b,c \in W$, $(a \cdot b) \cdot c = a \cdot (b \cdot c)$" is a statement of the _____ law of multiplication.

24. For all $a,b \in W$, $a \cdot b$ (always/sometimes/never) equals $b \cdot a$.

25. For all $a,b,c \in W$, $a \cdot (b + c)$ (always/sometimes/never) equals $a \cdot b + a \cdot c$.

26. If $a, b, c \in W$ and $a = b$, then $a \cdot c$ (always/sometimes/never) equals $b \cdot c$.

27. If $a, b \in W$, and if $a < b$, then $a - b$ is (always/sometimes/never) a whole number.

28. The difference $a - b$ is the number d (if it exists) such that $(a + d = b \ / \ b + d = a \ / \ b + a = d)$.

29. If $a, b \in W$, the quotient of a divided by b is (always/sometimes/never) a whole number.

30. The quotient $\frac{a}{b}$ is the number q (if it exists) such that $(b \cdot q = a \ / \ b \cdot a = q \ / \ a \cdot q = b)$.

If you missed fewer than eight questions on this test, continue on to Unit III of this program. If you missed eight or more questions, you would probably profit in returning to the remark on page 31 and reading through the program to this point before starting the next unit.

UNIT III

The Set of Integers

OBJECTIVES

Upon completion of this part of the program, the reader will be able to:

1. Use the correct vocabulary and/or the correct notation to express the concepts of an integer, additive inverse, ordering of integers, and absolute value.

2. Illustrate and identify the additive inverse, or negative, law.

3. Distinguish between a "negative number" and the "negative of a number."

4. Associate integers with points on a number line.

5. Use the order symbols $<$ and $>$, correctly in the ordering of any two elements in the set of integers.

6. Write the sum of two positive integers, two negative integers, and a positive and a negative integer as a basic numeral.

7. Use a number line to illustrate the addition of integers.

8. Write the product of two positive integers, two negative integers, and a positive and a negative integer as a basic numeral.

9. Express the difference of two integers a and b in terms of a sum.

10. Write the quotient of two integers, when such a quotient exists in the set of integers, as a basic numeral.

Remark. To this point in our discussion, our universe has been the set of whole numbers, $W = \{0, 1, 2, 3, \ldots\}$. Now we shall look at a set of numbers that contains all of the elements of W and some other elements besides. As you will see, this enlarged set has all the properties of the set of whole numbers, and also is closed with respect to subtraction. This new set of numbers, called the set of integers, will enable us to give meaning to a difference such as $4 - 7$, or, in general, $a - b$, for any ordered pair of whole numbers.

To govern the behavior of the new members of the enlarged set, we must adopt some new axioms. First, let us return briefly to a number line extended to the left of the origin which is marked off with segments of equal length:

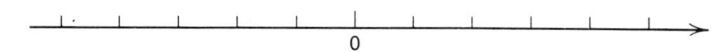

For each natural number corresponding to a point to the right of the origin, we shall assume that there exists a number that we can assign to a point located in the same relative position to the *left* of the origin.

1. Each of the numbers in a pair corresponding to points located the same number of units from the origin, but in opposite directions, is called the **negative** of the other. Thus, if the points C and D are located the same number of units from the origin on the number line

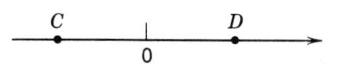

   then the number associated with the point C is the negative of the number associated with the point D, and similarly, the number associated with the point D is the _____ of the number associated with the point C.

   ---
   negative

2. The negative of a natural number is sometimes represented by a numeral that includes a raised dash. Thus, $^-2$ denotes the negative of 2, and is read "negative two." The symbol $^-3$ denotes the negative of 3. The symbol $^-5$ denotes the _____ of 5, and is read "_____ _____."

   ---
   negative; negative five

3. Numbers such as $^-3, ^-5, ^-11,$ and $^-337$ are referred to as **negative numbers**, or, in particular, as **negative integers**. Select the negative integers in the following set: $\{^-4, 7, 13, ^-9, ^-25, 19\}$.

   $^-4;\ ^-9;\ ^-25$

107

108 UNIT III

4. In general, to every natural number a, there corresponds a negative integer ^-a, the negative of a. Likewise, a is the _____ of ^-a.

---
negative

5. A one-to-one correspondence exists between the members of the set of natural numbers and the members of the set of negative _____ .

---
integers

6. Corresponding to the natural number 5 is its negative, _____ .

$^-5$

7. The graph of 6 appears as

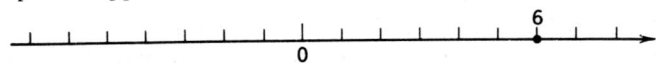

The negative of 6 is represented by the symbol _____ . Graph the negative of 6.

$^-6$;

8. The negative of $^-6$ is 6; the negative of $^-3$ is _____ .

3

9. Symbolically, the negative of $^-3$ can be written as $^-(^-3)$, and read "the negative of negative 3." But the negative of $^-3$ is 3, so that $^-(^-3) = 3$. Similarly, $^-(^-18) =$ _____ , and $^-($ _____ $) = 29$.

18; $^-29$

10. To distinguish between negative numbers and numbers (other than 0) that are not negative numbers, the term **positive number** is used. Any number corresponding to a point on the number line to the right of the origin is a _____ number.

---
positive

The Set of Integers 109

11. Any integer corresponding to a point on the number line to the right of the origin is a _____ integer.

positive

12. If a is a positive integer, then ^-a is a _____ integer.

negative

13. If a is a negative integer, then ^-a is a _____ integer.

positive

Remark. We have to distinguish between the two phrases "negative number" and "negative of a number." The phrase "negative number" refers to those numbers associated with points lying to the left of the origin. The phrase "negative of a number" refers to what we might call the "opposite" (in the sense of being on opposite sides of 0) of a number. The "opposite" of a negative number is associated with a point to the right of the origin, and we use the word "positive" to describe such numbers.

14. Occasionally the symbol + is prefixed to the symbol representing a positive number. Thus, $^+8$, $^+10$, $^+17$ represent _____ numbers.

positive

15. The symbols $^+3$ and 3 both represent the same _____ number, or _____ integer.

positive; positive (*Or natural; positive.*)

16. The symbols 7 and _____ represent the same positive number.

$^+7$

17. The set of all natural numbers and their negatives, together with the number 0, is called the **set of integers**. Thus, 3, $^-3$, 21, $^-21$, and 0 are integers. $^-3$ and $^-21$ are negative _____.

integers

18. The set of integers is a(n) (finite/infinite) set.

    infinite

19. All whole numbers are integers. Because 143 is a whole number, it is an
    _____ .

    integer

20. Not all integers are whole numbers. For example, ⁻18 (is/is not) a whole number; but ⁻18 (is/is not) an integer.

    is not; is

21. The number 0 (is/is not) an integer.

    is

Remark. We now need an axiom to govern the behavior of the negative numbers in the set of integers. Our reliance on the line graph in earlier frames was simply a device to introduce you to the negative integers. The following frames will characterize these numbers more formally. We shall take as an axiom the following:

For every integer a, there exists exactly one integer ⁻a, called the additive inverse of a, such that

$$a + {}^-a = 0 \text{ and } {}^-a + a = 0.$$

This statement is called the **Additive inverse law** or the **Negative law**.

22. The Additive inverse law asserts that $7 + {}^-7 = 0$, $29 + $ _____ $= 0$, and $32 + {}^-32 = $ _____ .

    ⁻29; 0

23. ⁻11 is the additive _____ of 11, and 11 is the _____ _____ of ⁻11.

    inverse; additive inverse

24. ⁻3 + 3 = _____ , 7 + _____ = 0, and _____ + 19 = 0.

    0; ⁻7; ⁻19

The Set of Integers

Remark. We have referred to a symbol such as ⁻a as either the negative of *a* or the additive inverse of *a*. These words are synonymous. We shall use the terms as it is convenient.

The relationship between positive and negative integers and their association with points on the number line are shown in the following figure.

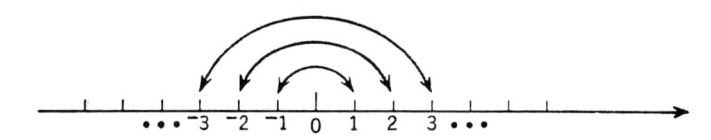

For convenience we shall designate the set of integers by the letter *J*. Hence
$$J = \{\ldots ^{-}3, ^{-}2, ^{-}1, 0, 1, 2, 3, \ldots\}.$$

25. Recall that the symbols $A \subset B$ mean that the set *A* is a subset of the set *B*. The set of whole numbers, $W = \{0, 1, 2, 3, \ldots\}$, is contained in the set of integers, $J = \{\ldots ^{-}2, ^{-}1, 0, 1, 2, 3, \ldots\}$. *W* is a subset of *J*, or symbolically, *W* _____ *J*.

$W \subset J$

26. As long as the symbols 0, 1, 2, 3, etc., refer to the same members in the set of whole numbers or the set of integers, then the number denoted by "2", for example, can be called:
    1. a natural number or a whole number;
    2. a positive _____; or
    3. the additive _____ of the integer ⁻2.

-----

integer; inverse

27. The number denoted by "⁻2" can be called:
    1. the _____ of a natural number or a whole number;
    2. a _____ integer; or
    3. the _____ inverse of the integer 2.

-----

negative; negative; additive

28. The number denoted by "5" can be called a natural number, a whole number, a _____ integer, or the _____ inverse of _____.

-----

positive; additive; ⁻5

UNIT III

29. 5 + ⁻5 = ____.

0

30. The number denoted by "⁻4" can be called the negative of a natural number, or a _____ integer, or the additive _____ of ____.

---
negative; inverse; 4 (Or ⁺4.)

31. ⁻4 + 4 = ____.

0

Remark. Because the numerals for negative integers involve the use of the "minus sign," for example ⁻4, and because positive integers are sometimes denoted by symbols like ⁺4, the positive and negative integers are sometimes called **signed numbers**.

We have observed that the whole numbers (zero and the positive integers) are *ordered*. Let us see how this ordering can be extended to include negative integers as well. To do this, we shall appeal again to the number line. Bear in mind, as we do so, that we are not trying to prove anything by using number lines. We use them because they give us something concrete to relate to abstract notions.

32. Recall that if a and b represent whole numbers, then a is either greater than b, equal to b, or less than b. If a is *less* than b, then the point corresponding to a on a number line will lie to the _____ of the point corresponding to b.

---
left

33. Because the point corresponding to a is to the *left* of the point corresponding to b on the number line

```
────|────●────────●─────▶
    0    a        b
```

the whole number represented by a is (less than/greater than) the whole number represented by b.

---
less than

The Set of Integers

34. The number line makes it plausible to assert that the integers are ordered in a manner consistent with the ordering of the whole numbers. That is, if a and b denote integers, and if the graph of a lies to the left of the graph of b,

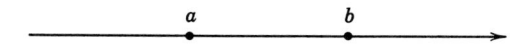

$$\xrightarrow{\qquad \underset{a}{\bullet} \qquad \underset{b}{\bullet} \qquad}$$

then we say that a is (less than/greater than) b.

---
less than

35. From the number line

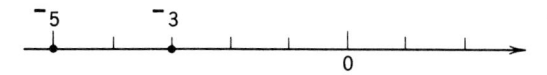

$$\xrightarrow{\underset{^-5}{\bullet} \quad \underset{^-3}{\bullet} \quad | \quad | \quad \underset{0}{|} \quad | \quad | \quad | \quad |}$$

it is evident that $^-5$ is (less than/greater than) $^-3$, because the graph of $^-5$ is to the left of the graph of $^-3$.

---
less than

36. Since every negative integer is located to the left of 0, and every positive integer is located to the right of 0, then every negative integer is (less than/greater than) every positive integer.

---
less than

37. Graph the numbers $^-8$, $^-3$, and 2 on the number line

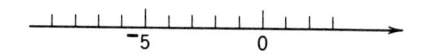

$$\xrightarrow{| \quad | \quad | \quad | \quad | \quad | \quad | \quad | \quad | \quad | \quad | \quad | \quad |}$$
$$\qquad\qquad\quad ^-5 \qquad 0$$

$$\xrightarrow{\underset{^-8}{\bullet} \quad | \quad | \quad \underset{^-3}{\bullet} \quad | \quad | \quad \underset{^-5 \quad\;\; 0}{\;} \quad \underset{2}{\bullet}}$$

38. From the preceding frame, we can say that $^-8$ is (less than/greater than) $^-3$, and 2 is (less than/greater than) $^-3$.

---
less than; greater than

Remark. You may have difficulty understanding that ⁻5, for example, is less than ⁻3, because you are accustomed to thinking about the set of whole numbers, wherein the symbol "5" denotes a number greater than that denoted by "3". One way to avoid confusion with regard to the relative order of two integers is to think of the number line when you want to consider whether one integer is greater than or less than another.

39. The relative order of the numbers ⁻8 and ⁻3 can be shown by writing ⁻8 < ⁻3. The relative order of ⁻5 and ⁻16 can be shown by writing ____ < ____ .

    ⁻16 < ⁻5

40. The relative order of the integers ⁻6 and 1 can be shown by writing ____ < ____ .

    ⁻6 < 1

41. ⁻6 < 1 means the same as 1 ____ ⁻6.

    1 > ⁻6

42. If a and b are integers and if $a < b$, then b ____ a.

    $b > a$

43. The relative order of the three numbers ⁻1, ⁻18, and 0 can be shown by writing ____ < ____ < ____ .

    ⁻18 < ⁻1 < 0

44. The relative order of the numbers ⁻7, 5, and ⁻13 can be shown as ____ < ____ < ____ .

    ⁻13 < ⁻7 < 5

Remark. At this point in this part of the program, we possess the set of integers $J = \{\ldots ⁻3, ⁻2, ⁻1, 0, 1, 2, 3, \ldots\}$, an axiom characterizing the additive inverse of each number in the set, and a way by which we can order the elements in the set.

The next short sequence of frames will give you an opportunity to review fundamental ideas that we have considered to this point that pertain to the system of integers.

The Set of Integers 115

R45. $J = \{\ldots {}^-3, {}^-2, {}^-1, 0, 1, 2, 3, \ldots\}$ is called the set of _____ .

integers

R46. $\{$Whole numbers$\}$ (is/is not) a subset of $\{$integers$\}$.

is

R47. The set of integers is a(n) (finite/infinite) set.

infinite

R48. "Natural numbers" and "_____ integers" refer to the same numbers.

positive

R49. For every integer a, there corresponds exactly one integer ${}^-a$ such that
$a + ({}^-a) =$ ____ .

$a + ({}^-a) = 0$

R50. ${}^-a$ is called the negative of a or the _____ inverse of a.

additive

R51. If a is a positive integer, ${}^-a$ is (always/sometimes/never) a negative integer.

always

R52. If a is a negative integer, ${}^-a$ is (always/sometimes/never) a negative integer.

never   (*If a is negative, ${}^-a$ is positive.*)

R53. The number denoted by "7" can be called a natural number, a whole number, a _____ integer, or the _____ _____ of ${}^-7$.

positive; additive inverse

R54. On the number line

the integer associated with point P is (less than/greater than) the integer associated with point Q.

---

less than

R55. Graph the numbers $^-6$, 0, 5 on the number line.

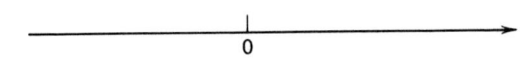

R56. The relative order of $^-5$, $^-9$, and 2 is ___ < ___ < ___ .

$^-9 < {^-5} < 2$

Remark. See Exercise IIIa, page 257, for additional practice on topics that have been considered to this point in this part of the program.

When a set of numbers is extended to form a new set, it is desirable that the properties assumed for the original set of numbers also can be applied in the new system. Therefore, we adopt all of the axioms for the set of whole numbers for the set of integers. (See Panel I, page 81, in Part II of this program.) These properties determine the nature of the system of integers. Because the axioms for the whole numbers were all that were necessary to prove some theorems about these numbers, and because these axioms are incorporated into the axioms for the system of integers, the theorems for the system of whole numbers are valid for the system of integers.

57. Recall that if $a, b, c \in W$ and if $a = b$, then $a + c = b +$ ___ , and by the Commutative law of addition, $c + a = c +$ ___ .

$a + c = b + c; \; c + a = c + b$

58. Recall that if $a, b, c \in W$ and if $a = b$, then $a \cdot c = b \cdot$ ___ , and by the Commutative law of multiplication, $c \cdot a = c \cdot$ ___ .

$a \cdot c = b \cdot c; \; c \cdot a = c \cdot b$

The Set of Integers 117

59. Thus, if $a = b$, then $a + 9 =$ _____ + _____ and $a \cdot 9 =$ _____ · _____ .

$a + 9 = b + 9;\ a \cdot 9 = b \cdot 9$

60. Because the axioms necessary to prove these theorems are incorporated in the axioms for the integers, their validity for the integers follows accordingly. Thus, if a, b, and c are *integers*, and if $a = b$, then $a + c =$ _____ and $a \cdot c =$ _____ .

$a + c = b + c;\ a \cdot c = b \cdot c$

61. If a and $b \in J$, and if $a = b$, then $a + (^-3) =$ _____, and $a \cdot (^-8) =$ _____ .

$b + (^-3);\ b \cdot (^-8)$

62. If $a \in J$, then $a \cdot 0 =$ _____ and $0 \cdot a =$ _____ .

$a \cdot 0 = 0;\ 0 \cdot a = 0$

Remark. Let us now consider operations on integers in more detail. We begin with the process of addition, considering first the sum of two positive integers, then the sum of two negative integers, and finally the sum of a positive integer and a negative integer.

63. Recall that the phrases "positive integers" and "natural _____" mean the same thing.

---

numbers

64. The sum of two positive integers will be the same as the _____ of two natural numbers. Both $3 + 4$ and $^+3 + {}^+4$ can be written as the basic numeral _____ .

---

sum; $^+7$ (*Or* 7.)

65. Because we have assumed that integers obey the Commutative law of addition, then $^+5 + {}^+8 = {}^+8 +$ _____ . Both of these representations for a sum can be written as the basic numeral _____ .

$^+8 + {}^+5;\ ^+13$ (*Or* $8 + 5$; 13.)

66. The foregoing frames suggest that the sum of two positive integers is a _____ integer.

---

positive

Remark. Because the sum of two natural numbers is always a natural number, and a positive integer is a natural number, we shall take the following statement as an axiom.

For all $a, b \in J$, $a, b > 0$, $a + b$ is a positive integer.

Recall that for a, b, and c, elements in the set of *whole numbers*, we defined $a + b + c$ as $(a + b) + c$. We shall define $a + b + c$ in the same way when a, b, and c are *integers*.

The sum of two positive integers is a positive integer. Now, we shall argue that the sum of the negatives of two integers will be negative. We first consider a specific case and show that $(^-3 + {^-5}) = {^-(3 + 5)} = {^-8}$.

67. Since $3 + {^-3} = 0$ and $5 + {^-5} = 0$ by virtue of the Additive inverse law, $(3 + {^-3}) + (5 + {^-5}) = $ _____ .

0

68. The associative and commutative laws justify writing $(3 + {^-3}) + (5 + {^-5}) = 0$ as

$$(3 + 5) + (^-3 + {^-5}) = 0.$$

Hence, $(^-3 + {^-5})$ must be the additive inverse of $(3 + 5)$. That is

$$(^-3 + {^-5}) = {^-(\quad + \quad)}.$$

$^-(3 + 5)$

Remark. The preceding sequence of frames suggests the following theorem.

Theorem: If $a, b \in J$, then $^-a + {^-b} = {^-(a + b)}$.

Proof:

$0 + 0 = 0$	(Identity law of addition)
$a + {^-a} = 0$; $b + {^-b} = 0$	(Additive inverse law)
$(a + {^-a}) + (b + {^-b}) = 0$	(Substitution law)
$(a + b) + \boxed{(^-a + {^-b})} = 0$	(Associative and Commutative laws of addition)

Since the additive inverse of $(a + b)$ is also given by $^-(a + b)$, we have that

$\boxed{^-a + {^-b}} = {^-(a + b)}$ (uniqueness of an additive inverse)

The Set of Integers   119

The uniqueness of an additive inverse (part of the Additive inverse law) implies that there is exactly *one* number with the property that the sum of this number and a second number equals zero.

The above proof is given in detail to give you some notion of how an algebraic proof can be constructed. However, we will only need the last statement in the work that follows.

As a special case of the result $^-a + {}^-b = {}^-(a + b)$, if we restrict a and b to be positive integers, then ^-a and ^-b are negative integers, and the theorem says that *the sum of two negative integers is a negative integer.* Let us apply this theorem a few times to see how it works.

69. $^-3 + {}^-8 = {}^-(\underline{} + \underline{}) = \underline{}$.

$^-(3 + 8)$; $^-11$

70. $^-14 + {}^-7 = \underline{}$.

$^-21$

71. The expression $^-2 + {}^-6$ is read "negative two plus negative six," and represents the _____ of $^-2$ and $^-6$. $^-2 + {}^-6$ can be written as the basic numeral _____ .

---

sum; $^-8$

72. $^-25 + {}^-3 = \underline{}$.

$^-28$

73. $^-3 + {}^-9 = {}^-9 + {}^-3$ is an example of the application of the _____ law of addition for the set of integers.

Commutative

74. Recall that the Associative law of addition is assumed to hold for integers. For example, $({}^-2 + {}^-3) + {}^-4$ is the same as $^-2 + (\underline{} + \underline{})$. Each sum can be represented by the basic numeral _____ .

$({}^-3 + {}^-4)$; $^-9$

120 UNIT III

75. $(^-5 + {}^-2) + {}^-3 = {}^-5 + ({}^-2 + {}^-3)$ is an example of the _____ law of addition. Each sum can be written as the basic numeral _____.

Associative; $^-10$

76. $(^+2 + {}^+4) + {}^+7 = {}^+2 + (\underline{} + \underline{})$. Each sum can be written as the basic numeral _____.

$(^+4 + {}^+7)$; 13

77. $^-5 + 0 + {}^-22 =$ _____, and $0 + {}^+8 + {}^+6 =$ _____.

$^-27$; 14

Remark. To this point in our discussion of the addition of integers, we have restricted ourselves to the addition of two positive integers or the addition of two negative integers. You have seen that *when we add a positive number to a positive number, we obtain a positive number and when we add a negative number to a negative number, we obtain a negative number.* When we add two numbers, however, one of which is negative and the other positive, the sum *may be positive* or it *may be negative*, depending upon the numbers being added.

To prepare for a discussion of this kind of sum in the set of integers, we shall first introduce the expression "absolute value," and the corresponding symbolic notation.

78. The numbers 4 and $^-4$ correspond to points located the same number of units from the origin on a number line, but in opposite directions. To refer only to the number of units, and not to the direction, the symbols $|^-4|$ and $|4|$ are used. The symbol $|^-4|$ is read "the **absolute value** of negative four," and corresponds to a point 4 units from the origin, without regard to direction. In a similar way $|^-2|$ is read "the _____value of negative two" and indicates that the point is ____ units from the origin.

---
absolute; 2

79. $|4|$ is read "the absolute value of four" and corresponds to a point 4 units from the origin. $|8|$ is read "the _____ value of eight."

---
absolute

The Set of Integers

80. Both $|^-9|$ and $|9|$ correspond to points located the same number of units from the origin. Thus, $|^-9|$ is (less than/equal to/greater than) $|9|$.

    ---
    equal to

81. The absolute value of 0, or $|0|$, is 0. The absolute value of any number other than 0 is positive. Thus, $|4| = 4$, $|0| = $ _____, and $|^-4| = $ _____.

    0; 4

82. $|7|$ (is/is not) equal to $|^-7|$.

    ---
    is

Remark. Now see how we can employ the idea of absolute value to consider the sum of a positive and a negative integer.

83. Consider the sum $^+3$ and $^-4$. By judicious choice, $^+3$ and $^-4$ can be written $^+3 + (^-3 + {}^-1)$. By the Associative law of addition, $^+3 + (^-3 + {}^-1) = (^+3 + {}^-3) + $ _____.

    $^-1$

84. Since $^+3 + {}^-4 = {}^+3 + (^-3 + {}^-1)$,
    then $^+3 + {}^-4 = (^+3 + {}^-3) + {}^-1$
    $= 0 + {}^-1 = $ _____.

    $^-1$

Remark. In the above example, the sum of $^+3$ and $^-4$ was shown to be equal to $^-1$. Hence, in this case, the sum of a positive integer and a negative integer was a *negative* integer. Notice further that in this example, $|^-4| > |3|$. But this is not always the situation. Let us consider the sum $^-7 + {}^+9$.

85. The sum $^-7 + {}^+9$ can be rewritten $^-7 + (^+7 + {}^+2)$. By the Associative law of addition, we can write $^-7 + (^+7 + {}^+2) = (^-7 + {}^+7) + $ _____.

    $^+2$ (*Or* 2.)

121

122   UNIT III

86. Since $-7 + {}^+9 = -7 + ({}^+7 + {}^+2)$,
  then $-7 + {}^+9 = (-7 + {}^+7) + {}^+2$
  $\phantom{-7 + {}^+9\ } = 0\phantom{-7 + {}^+7)} + {}^+2 = $ _____.

${}^+2$ (Or 2.)

Remark. So, $-7 + {}^+9 = {}^+2$. In this case, the sum of a positive integer and a negative integer is a positive integer. Note in this example that $|9| > |{}^-7|$. Is it clear to you what is happening? *The sum of a positive integer and a negative integer will be positive or negative, depending upon which of the integers in a sum has the greater absolute value. In fact, the sum will be the difference between the absolute values, and will be positive if the positive integer has the greater absolute value, and negative if the negative integer has the greater absolute value.*

The preceding sequence of frames suggest the following theorem:

  If $a, b \in J$, and
      if $a > b > 0$, then $a + (-b) = {}^+(a - b)$;
      if $b > a > 0$, then $a + (-b) = {}^-(b - a)$.

We shall not demonstrate a formal proof of this theorem. However, you should be able to apply this theorem and other properties that have been introduced concerning the sums of integers.

87. The sum $5 + (-3) = {}^+(5 - 3) = $ _____ and the sum $-5 + 3 = {}^-(5 - 3) = $ _____.

$2; {}^-2$

88. The sum $-4 + 7 = {}^+(7 - 4) = $ _____ and the sum $4 + (-7) = {}^-(7 - 4) = $ _____.

$3; {}^-3$

89. The sum $9 + 5 = $ _____ , $-9 + (-5) = $ _____ , $-9 + 5 = $ _____ and $9 + (-5) = $ _____.

$14; {}^-14; {}^-4; 4$

90. The sum $9 + 0 = $ _____ and the sum $0 + {}^-5 = $ _____.

$9; {}^-5$

Remark. As you will see in the following sequence of frames, the number line can be helpful in visualizing the operation of addition in the set of integers. However you

The Set of Integers 123

should understand that the results of the operation are basically determined by the definitions we have made and the axioms that we have adopted, and not by the representation of the operation on the number line.

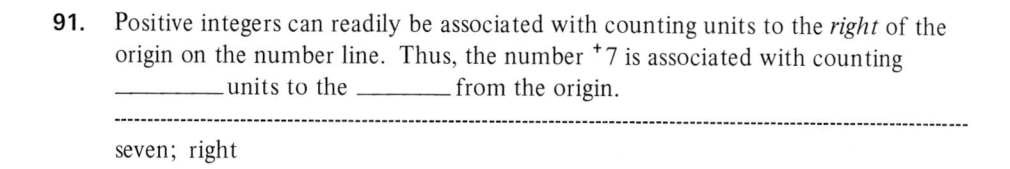

91. Positive integers can readily be associated with counting units to the *right* of the origin on the number line. Thus, the number $^+7$ is associated with counting _____ units to the _____ from the origin.

---
seven; right

92. The sum $^+4 + {}^+2$ can be thought of as the number associated with the point arrived at after counting four units to the right, starting from the origin, and then, from this point, counting two more units to the right, as shown on the number line.

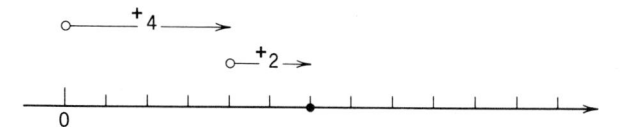

The number associated with the point reached last is _____. The sum of $^+4$ and $^+2$ is _____.

$^+6$; $^+6$ (*Or,* $^+4 + {}^+2$. *The symbols $^+6$ and $^+4 + {}^+2$ represent the same sum. $^+6$ is the basic numeral.*)

93. Show that the graphing process produces the same result if $^+4 + {}^+2$ is written $^+2 + {}^+4$. Repeat the process used in the preceding frame, but this time begin by counting two units to the right from the origin.

This shows that graphical addition also obeys the (Closure/Commutative/Associative) law of addition.

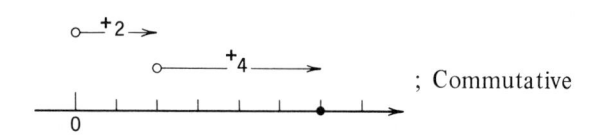

; Commutative

124 UNIT III

94. Negative integers can be associated with counting to the *left* on a number line. Thus, the number ⁻3 is associated with counting three units to the _____ from the origin.

---
left

95. The sum of ⁻5 and ⁻3 can be found by first counting five units to the left from the origin, and, from this point, counting three more units to the left. On the number line below, the number associated with the point reached last is _____ .

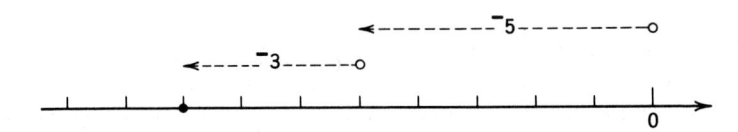

The sum of ⁻5 and ⁻3 is _____ .

⁻8; ⁻8 (Or ⁻5 + ⁻3.)

96. To illustrate commutativity in the sum of two negative integers, show the addition of ⁻3 and ⁻5 on the number line, graphing ⁻3 first.

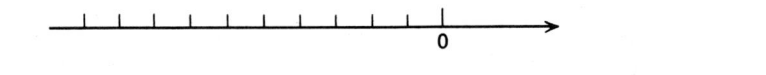

---

97. The previous examples, in which we used number lines, support the consequences of our axioms in that the sum of two positive integers is a positive integer and that the sum of two negative integers is a _____ integer.

---
negative

Remark. Now we shall see how the sum of any positive and any negative integer can be visualized on the number line.

The Set of Integers

98. The sum $^+7 + {}^-5$, for example, can be thought of as the number associated with the point reached last, after counting seven units to the right from the origin and then, from this point, counting five units to the left, as shown.

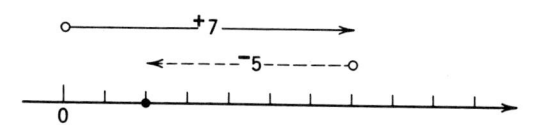

The number associated with the point reached last in this process is _____. The sum of $^+7$ and $^-5$ is _____.

$^+2$; $^+2$ (Or $^+7 + {}^-5$.)

99. Since $^+7 + {}^-5 = {}^+2$, then, by the Commutative law of addition, $^-5 + {}^+7 =$ _____.

$^+2$

100. Show $^-5 + {}^+7$ on the number line, beginning by counting to the left.

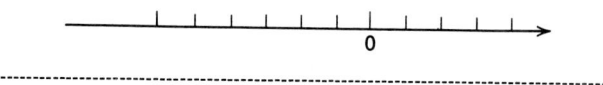

---

101. If we wish to find the sum of $^+3$ and $^-8$, three units are counted off to the right from the origin, and, from this point, eight units are counted off to the left.

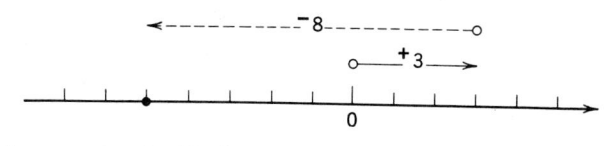

The number associated with the point reached last is _____. The sum of $^+3$ and $^-8$ is _____.

$^-5$; $^-5$

102. Use the number line below to show that $^-8 + {}^+3$ is also equal to $^-5$. Begin by counting to the left.

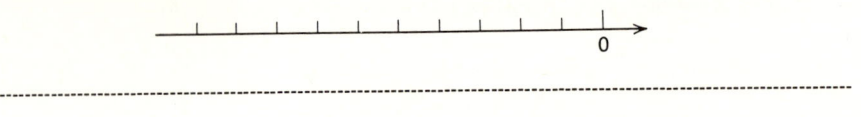

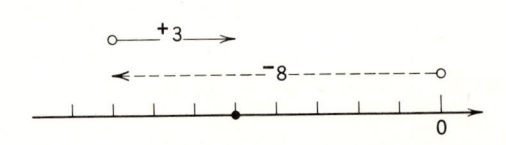

Remark. We hope that this brief treatment of sums of integers using the number line makes you feel more at ease with the process of the addition of integers. The next few frames will give you some practice writing sums as basic numerals. If necessary, think of the number line.

103. $^-3 + {}^+2 =$ _____ ; $^+5 + {}^-4 =$ _____ .

$^-1$; $^+1$

104. $^-5 + {}^+8 =$ _____ ; $^+3 + {}^-19 =$ _____ ; $^-6 + {}^+6 =$ _____ .

$^+3$; $^-16$; 0

Remark. The numerals we have been using to represent integers, that is, numerals such as $^-8$ and $^+3$, are not widely employed for the purpose. The signs that are part of the numerals are normally centered, rather than appearing as superscripts. That is, $^-8$ is generally denoted by -8, and $^+3$ is written $+3$, or, simply 3. If this symbolism is adopted, however, it becomes necessary to use parentheses in denoting sums, because, where $^-8 + {}^+3$ is clear, $-8 + +3$ is not clear. Therefore, numerals for integers are frequently enclosed in parentheses to keep the sign portion of the numeral distinct from the symbol used to denote an operation. For example, $^-8 + {}^+3$ might be written $(-8) + (+3)$ or $(-8) + (3)$. Indeed, since no ambiguity is possible with respect to $^-8$ or 3, it might be written $-8 + 3$.

105. $-9 + (-6)$ means $^-9 + {}^-6$. $-8 + (-7)$ means _____ + _____ .

$^-8 + {}^-7$

The Set of Integers

106. ⁻3 + ⁻6 could be written using the numerals −3 and −6 as ____ + ____ .

(−3) + (−6)   (Or −3 + (−6).)

Remark. Recall that when we were studying the whole numbers, we were unable to give meaning to an expression such as 9 − 12. We said, therefore, that the set of whole numbers was not closed with respect to the operation of subtraction. Now, however, we define the difference of two integers in such fashion that the set of integers is closed and the definition is consistent with the definition for a difference in the set of whole numbers.

$$\text{For all } a,b \in J, \ a - b = a + (-b).$$

This definition can be shown (although we shall not do so here) to be consistent with the definition of a difference in the set of whole numbers on page 85 of Part II of this program.

An expression such as 9 − 12 means 9 + (−12) for which the basic numeral is −3. Similarly, 12 − 9 means 12 + (−9) for which the basic numeral is 3. This latter example shows that the results obtained subtracting integers is consistent with the results we obtained in the system of natural numbers. Furthermore, every ordered pair of whole numbers has a difference in the set of integers.

In the set of integers, we need not be overly concerned with the operation of subtraction since *the difference of two numbers is equal to the sum of the first and the additive inverse of the second.*

107. 7 − 11 denotes the sum of +7 and −11. This sum can be written as the basic numeral _____ .

−4   (Or ⁻4. Hereafter, we will use −a in preference to ⁻a.)

108. +7 − 9 denotes the _____ of +7 and −9 and can be written as the basic numeral _____ .

---

sum; −2

109. Generally, if the positive number is the first of the two numbers in a sum, the "+" part of the numeral is omitted. Thus, 8 − 12 denotes the _____ of +8 and −12. The sum can be written as the basic numeral _____ .

---

sum; −4

128   UNIT III

110. 15 − 3 denotes the sum of +15 and _____. The sum can be written as the basic numeral _____.

−3; 12

Remark. The next frames will provide you with some practice writing sums of integers as basic numerals. If you have difficulty thinking of each representation as a sum, rewrite the expression using parentheses.

111. 8 − 6 = _____ ; 25 − 29 = _____.

2; −4   [(+8) + (−6) = 2; (+25) + (−29) = −4.]

112. −7 + 16 = _____ ; 0 − 8 = _____.

9; −8   [(−7) + (+16) = 9; (0) + (−8) = −8.]

Remark. The following frames will provide you with some more practice in determining basic numerals for sums. Since three numbers are involved, use the definition

$$a + b + c = (a + b) + c$$

and the associative law as you see fit.

113. 3 − 5 + 8 = _____.

---
6   [(3 − 5) + 8 = −2 + 8 = 6.]

114. −1 − 2 − 3 = _____ ; −3 + 9 − 7 = _____.

−6; −1

115. 0 − 7 + 7 = _____ ; 8 − 16 + 8 = _____.

0; 0

Remark. This concludes our discussion of sums and differences in the system of integers. The next sequence of frames will review these ideas.

The Set of Integers 129

R116. The statement that the sum of two integers is always an integer is justified by the (Closure/Commutative/Associative) law for addition.

Closure

R117. The sum of two positive integers is (always/sometimes/never) a positive integer.

always

R118. The sum of two negative integers is (always/sometimes/never) a negative integer.

always

R119. $6 + 3 + 0 =$ ____ ; $-5 + 0 - 22 =$ ____ .

9; −27

R120. $|-13|$ is read "the _____ value of negative thirteen" and can be written as the basic numeral ____ .

absolute; 13

R121. $|-8|$ (is/is not) equal to $|8|$.

is

R122. The sum of a positive integer and a negative integer is (always/sometimes/never) a negative integer.

sometimes (*It depends on which absolute value is greater.*)

R123. $-9 + 8 - 5 =$ ____ . $-6 + 9 + 2 =$ ____ .

−6; 5

R124. The difference of two integers a and b is defined as follows: $a - b = a + ($ ___ $)$.

$a - b = a + (-b)$

R125. $7 - 4 = 7 + ($ ___ $) =$ ___ . $4 - 7 = 4 + ($ ___ $) =$ ___ .

$7 + (-4) = 3$; $4 + (-7) = -3$

Remark. See Exercise IIIb, page 259, for additional practice on topics concerning sums and differences of integers.

Next, we consider the multiplication of integers. We want our results to be consistent with those obtained in the set of whole numbers and the axioms we have taken to govern operations in the set of integers. (See Panel I on page 81.)

We first consider the product of two positive integers. Because the product of two natural numbers is always a natural number, and a positive integer is a natural number, we shall adopt the following axiom:

> For all $a,b \in J$, $a,b > 0$, $a \cdot b$ is a positive integer.

The next sequence of frames will develop the idea that *the product of a positive integer and a negative integer is a negative integer.*

126. If a and b are integers, and if $a + b = 0$, then b must be equal to $-a$. That is, b must be the negative of a. Thus, if $3 + (b) = 0$, then $b =$ ___ .

-3

127. The same idea is true if the terms are products of two factors. Thus, if

$$(4) \cdot (3) + (a) \cdot (b) = 0,$$

then $(a) \cdot (b)$ must be the _____ of $(4) \cdot (3)$.

------

negative

128. By the distributive law,

$$4 \cdot [3 + (-3)] = 4 \cdot (3) + 4 \cdot (-3).$$

Since $3 + (-3) = 0$, the product $4 \cdot [3 + (-3)]$ equals 0, and

$$4 \cdot (3) + 4 \cdot (-3) = \underline{}.$$

0

The Set of Integers

129. $4 \cdot (3) + 4 \cdot (-3) = 0$ only if $4 \cdot (-3)$ is the _____ of $4 \cdot (3)$.

---

negative

130. Since $4 \cdot (-3)$ is the negative of $4 \cdot (3)$, and since $4 \cdot (3)$ is 12, the product $4 \cdot (-3)$ must equal _____.

-12

Remark. The previous sequence of frames suggests the theorem:

Theorem: If $a, b \in J$, then $(a) \cdot (-b) = -(a \cdot b)$.

Proof:

$b + (-b) = 0$	(Additive inverse law)
$a \cdot [b + (-b)] = a \cdot 0$	(Theorem: if $a = b$, then $c \cdot a = c \cdot b$)
$a \cdot b + a \cdot (-b) = a \cdot 0$	(Distributive law)
$a \cdot b + a \cdot (-b) = 0$	(Theorem: $a \cdot 0 = 0$)
$a \cdot (-b) = -(a \cdot b)$	(Additive inverse law)

This concludes our argument. Now, by restricting a and b to denote positive integers, we have that *the product of a positive integer and a negative integer is a negative integer.*

131. From the theorem in the previous remark, we have that $2 \cdot (-3) = -(2 \cdot 3) = -6$. Similarly, $3 \cdot (-5) = -(3 \cdot 5) =$ _____.

-15

132. By the Commutative law of multiplication, $(5) \cdot (-6) = (-6) \cdot (5) = -30$. Similarly, $(-6) \cdot (11) = (11) \cdot (-6) =$ _____.

-66

133. $(-6) \cdot (13) =$ _____, and $(9) \cdot (-8) =$ _____.

-78; -72

Remark. Now we shall consider the product of two negative integers.

134. If
$$(-3) \cdot (4) + (a \cdot b) = 0,$$
then $a \cdot b$ must equal 12, the negative of -12.

▼

If
$$(-3) \cdot (5) + (a \cdot b) = 0,$$
then $a \cdot b =$ _____ , the negative of $(-3) \cdot (5)$.

15

135. By the Distributive law,
$$-3 \cdot [5 + (-5)] = (-3) \cdot (5) + (-3) \cdot (-5).$$
But, since $-3 \cdot [5 + (-5)]$ also equals $-3 \cdot (0)$ or 0, then the expression
$$(-3) \cdot (5) + (-3) \cdot (-5) = \underline{}.$$

0

136. If
$$(-3) \cdot (5) + (-3) \cdot (-5) = 0,$$
then the product $(-3) \cdot (-5)$ must be the negative of the product $(-3) \cdot (5)$. Thus, $(-3) \cdot (-5) = 15$. Similarly, since $(-4) \cdot (5) = -20$, then $(-4) \cdot (-5)$ must be the negative of -20 and can be written as the basic numeral _____ .

20

Remark. The previous sequence of frames suggest that the product of two negative integers is a positive integer, and so it is. We have the theorem:

Theorem: If $a,b \in J$, then $(-a) \cdot (-b) = a \cdot b$

Proof:

$b + (-b) = 0$	(Additive inverse law)
$(-a) \cdot [b + (-b)] = -a \cdot 0$	(Theorem: if $a = b$, then $c \cdot a = c \cdot b$)
$(-a) \cdot (b) + (-a) \cdot (-b) = -a \cdot 0$	(Distributive law)
$(-a) \cdot (b) + (-a) \cdot (-b) = 0$	(Theorem: $a \cdot 0 = 0$)
$(-a) \cdot (-b) = -[(-a) \cdot (b)]$	(Additive inverse law)
$(-a) \cdot (-b) = -[-(a \cdot b)]$	(Theorem: $(-a) \cdot (b) = -(a \cdot b)$)

Since $-[-(a \cdot b)] = a \cdot b$, it follows that
$$(-a) \cdot (-b) = a \cdot b$$

This establishes our theorem. Again, if we now restrict a and b to denote positive integers, then we have that *the product of two negative integers is a positive integer.*

The Set of Integers 133

We have now covered all of the possibilities for the products of two integers, and these can be stated as follows:

1. The product of two positive integers is a positive integer;
2. The product of a positive integer and a negative integer is a negative integer;
3. The product of two negative integers is a positive integer; and
4. The product of any integer and 0 is 0.

137. $(+7) \cdot (+4) =$ _____ ; $(-3) \cdot (-8) =$ _____ .

28; 24

138. $(+1) \cdot (-7) =$ _____ ; $(-5) \cdot (+3) =$ _____ .

−7; −15

139. $(-3) \cdot (0) =$ _____ ; $(0) \cdot (+2) =$ _____ .

0; 0

Remark. We can apply these facts about the products of two integers (sometimes called "laws of signs") to products involving more than two factors. As we did in the similar case of addition, we shall simply define $a \cdot b \cdot c$ as follows:

If $a,b,c \in J$, then $a \cdot b \cdot c = (a \cdot b) \cdot c$.

140. Recall that the Associative law of multiplication for integers states that:

If $a,b,c \in J$, $(a \cdot b) \cdot c = a \cdot (b \cdot c)$.

Thus, $(2) \cdot (3) \cdot (4)$ means the same as $[(2) \cdot (3)] \cdot (4)$ or $(2) \cdot [(3) \cdot (4)]$. Both expressions can be written as the basic numeral _____ .

24

141. Write $(-5) \cdot (-1) \cdot (3)$ as a basic numeral.

15

142. Write $(2) \cdot (-4) \cdot (-3)$ as a basic numeral.

24

143. $(2) \cdot (3) \cdot (-6) = $ _____ .

-36

144. If 0 is a factor of any product, then the product must be equal to 0. For example, $(3) \cdot (0) \cdot (2) = 0$ and $(7) \cdot (-4) \cdot (0) = $ _____ .

0

Remark. This brings us to the end of our discussion of multiplication in the system of integers. The next sequence of frames will review a few ideas about products of integers.

R145. The statement that the product of two integers is always an integer is justified by the _____ law for multiplication.

Closure

R146. The product of two positive integers is (always/sometimes/never) a positive integer.

----

always

R147. The product of a positive integer and a negative integer is (always/sometimes/never) a positive integer.

----

never

R148. The product of two negative integers is (always/sometimes/never) a positive integer.

----

always

R149. $(3) \cdot (5) = $ _____ . $(-2) \cdot (5) = $ _____ . $(-6) \cdot (-11) = $ _____ . $(-3) \cdot (0) = $ _____ .

15; -10; 66; 0

Remark. Panel I of Unit II of this program includes the axioms we adopted for the set of whole numbers and some properties which were consequences of these axioms. Panel II, which follows, contains the axioms and properties, introduced in this part of the program as we extended the set of whole numbers to form the set of integers. You may want to refer to this panel as well as Panel I as you complete the remainder of the program.

Panel II

$$a, b \in J$$

AXIOMS

$a + (-a) = 0$ and $-a + a = 0$. **Additive inverse law** or **Negative law**

For all $a, b > 0$, $a + b > 0$ and $a \cdot b > 0$.

PROPERTIES

$-a + (-b) = -(a + b)$.

If $a > b > 0$, then $a + (-b) = +(a - b)$.

If $b > a > 0$, then $a + (-b) = -(b - a)$.

$a - b = a + (-b)$.

$(a)(-b) = -(a \cdot b)$.

$(-a)(-b) = a \cdot b$.

The Set of Integers 137

Remark. We shall now consider the quotient of two integers.

150. Recall that in the system of whole numbers we defined a quotient as a factor in a product. Formally, we stated:

For all $a, b \in W$, $\frac{a}{b}$ is the whole number q (if one exists) such that $b \cdot q = a$.

Thus, $\frac{8}{4}$ is the number q such that $4 \cdot q = 8$. Since $4 \cdot 2 = 8$, the quotient of 8 divided by 4 is 2. Similarly, $\frac{12}{4} = 3$, because $4 \cdot 3 = 12$. The quotient $\frac{18}{3} = $ _____, because _____ · _____ = _____.

6; $3 \cdot 6 = 18$

151. The quotient of two whole numbers is not always a whole number. For example, $\frac{3}{4}$ is not a whole number, because there is no whole number q such that $4 \cdot q = 3$. Thus, the set of whole numbers is not closed with respect to _____.

---

division

Remark. Now we want a definition of division for the set of integers that will be consistent with that of a quotient in the set of whole numbers. Hence, we adopt the following definition:

For all $a, b \in J$, $\frac{a}{b}$ is the integer q (if one exists) such that $b \cdot q = a$.

152. For example, $\frac{10}{-2} = -5$, because $(-2) \cdot ($ _____ $) = $ _____.

$(-2) \cdot (-5) = 10$

153. The quotient, $\frac{-3}{7}$ is not an integer because no integer, q, exists such that $7 \cdot q = $ _____.

-3

138   UNIT III

154. If a and b are integers ($b \neq 0$), the quotient $\dfrac{a}{b}$ is an integer *only* if b is a factor of a.

Therefore, the set of integers (is/is not) closed with respect to the operation of division.

---
is not

Remark. Note that while the set of integers was closed for subtraction, this is not true with respect to the operation of division. A quotient of the form $\dfrac{a}{b}$, a and b integers, and $b \neq 0$, is defined in the set of integers if b is a factor of a; the quotient is not defined if b is not a factor of a. In the next part of the program, we will add new members to the set of integers to form a more extensive set of numbers in which quotients will be included where b is not a factor of a.

See Exercise IIIc, page 262, for additional practice on topics of multiplication and division in the set of integers.

Remark. Let us now review what you have learned about integers in this entire unit.

R155. By including the negatives of the natural numbers in the set of whole numbers, a new set, $J = \{\ldots {}^-3, {}^-2, {}^-1, 0, 1, 2, 3 \ldots\}$, the set of _____, was formed. This set is (finite/infinite).

---
integers; infinite

R156. An axiom was adopted to govern the behavior of the new members of this enlarged set.

> For every $a \in J$, there exists exactly one integer ${}^-a$, such that $a + {}^-a = $ ____ and ${}^-a + a = $ ____ .

This axiom is called the _____ inverse law or the Negative law.

0; 0; Additive

R157. ${}^-a$ is called the negative or _____ _____ of a. Likewise, a is the negative of ____.

---
additive inverse; ${}^-a$

The Set of Integers

R158. The negative of ⁻7 is _____ .

7

R159. The set of integers contains the natural numbers, the negatives of the natural numbers, and _____ .

0

R160. If a is a negative integer, then ⁻a is a natural number or a _____ integer.

---
positive

R161. The integers are ordered in a manner consistent with the ordering of the whole numbers. Thus, on the number line

```
            P   Q
────────┼───•───•──────▶
        0
```

the integer represented by point P is (less than/greater than) the integer represented by point Q.

---
less than

R162. On the number line

```
     S    R
─────•────•───┼──────▶
              0
```

the integer represented by point R is (less than/greater than) the integer represented by point S.

---
greater than

R163. ⁻7 is (less than/greater than) ⁻4, and 2 is (less than/greater than) ⁻4.

---
less than; greater than

R164. The axioms pertaining to the whole numbers (are/are not) adopted as axioms for the integers.

---
are

R165. The validity of theorems proven for the set of whole numbers also follows for the set of integers. One of these theorems states:

If $a, b, c \in J$, and if $a = b$, then $\underline{a + c =}$ _____ $+$ _____ .

$a + c = b + c$

R166. A second theorem states:

If $a, b, c \in J$ and if $a = b$, then $\underline{a \cdot c =}$ _____ .

$a \cdot c = b \cdot c$

R167. A third theorem states:

For each $a \in J$, $a \cdot 0 =$ _____ .

$a \cdot 0 = 0$

R168. Since the phrases "positive integers" and "natural numbers" mean the same thing, and since the sum of two natural numbers is always a natural number, it follows that:

If a and b are positive integers, then $a + b$ is a _____ integer.

---

positive

R169. The sum of two negative integers is always a negative integer. Thus:

If a and b are positive integers, then $^-a + {^-b} = {^-}($ _____ $)$.

$^-a + {^-b} = {^-}(a + b)$

R170. The absolute value of any nonzero integer is positive. Hence, $|^-4|$ can be written as the basic numeral _____ .

4

R171. If a is an integer, $|^-a|$ (is/is not) equal to $|a|$.

---

is

The Set of Integers 141

R172. The sum of a positive integer and a negative integer is the difference between the absolute values and is *positive* if the absolute value of the positive integer is (greater than/less than) the absolute value of the negative integer.

---
greater than

R173. The sum of a positive integer and a negative integer is *negative* if the absolute value of the positive integer is (greater than/less than) the absolute value of the negative integer.

---
less than

R174. If $a, b \in J$ and if $a > b > 0$, then $a + (-b) = {}^+($ _____).

$a + (-b) = {}^+(a - b)$

R175. If $a, b \in J$ and if $b > a > 0$, then $a + (-b) = {}^-($ _____).

$a + (-b) = {}^-(b - a)$

R176. ${}^+7 + {}^-5$ can be written as the basic numeral _____ ; ${}^-8 + {}^+4$ can be written as the basic numeral _____ .

2; ${}^-4$

R177. The signs that are part of the numerals, such as ${}^-8$ and ${}^+3$, are normally centered, rather than appearing as superscripts. Thus, ${}^-8$ is denoted _____ and ${}^+3$ is written _____ .

−8; +3 (Or −8; 3.)

R178. The sum ${}^-8 + {}^+3$ can be written $(-8) + (+3)$, or without parentheses as _____ + _____ . $+7 + (-9)$ can be written without parentheses as $+7 - 9$, where it is understood that $+7$ and _____ are to be added.

−8 + 3; −9

142 UNIT III

R179.   If the positive number is the first of the two numbers in a sum, the "+" part of the numeral is omitted. Thus, 8 − 12 denotes the _____ of +8 and −12, and can be written as the basic numeral _____.

---

sum; −4

R180.   Consistent with our definition of a difference in the system of whole numbers, we have that for all $a, b \in J$, $a - b = a + (\;\;\;\;)$.

$a - b = a + (-b)$

R181.   Thus, $13 - 7 = 13 + (-7) =$ ___, and $7 - 13 = 7 + ($ ___$) =$ ___.

6; −13; −6

R182.   We can view the symbols $7 - 13$ as the _____ of 13 subtracted from 7 or as the _____ of 7 and −13.

---

difference; sum

R183.   Since the difference of two integers is defined in terms of a sum, such a difference is always an integer. Therefore the set of integers (is/is not) closed for subtraction.

---

is

R184.   Write $^+4 + {^-8} + {^+3}$ as a basic numeral.

$^-1$

R185.   The product of two positive integers is a _____ integer; the product of a positive integer and a negative integer is a _____ integer; the product of two negative integers is a _____ integer; the product of any integer and zero is _____.

---

positive; negative; positive; zero

R186.   The product $(2) \cdot (6)$ can be written as the basic numeral _____. Also, $(2) \cdot (-6) =$ _____, $(-2) \cdot (-6) =$ _____, and $(0) \cdot (-6) =$ _____.

12; −12; 12; 0

The Set of Integers

R187. Consistent with the definition for the quotient of two whole numbers, we have that for all $a, b \in J$, $\dfrac{a}{b}$ is the integer q (if one exists) such that ____ $\cdot q =$ ____ .

$b \cdot q = a$

R188. Because an integer may not always exist for $\dfrac{a}{b}$, the set of integers (is/is not) closed with respect to division.

---

is not

R189. The symbol $\dfrac{-8}{4}$ represents an integer, -2, because $4 \cdot (-2) =$ ____ . The symbol $\dfrac{-7}{4}$ (does/does not) represent an integer.

-8; does not

R190. $\dfrac{a}{0}$ $(a \neq 0)$ (does/does not) represent a member of the set of integers because there is *no* integer q such that $0 \cdot q = a$.

---

does not

R191. $\dfrac{0}{b}$ $(b \neq 0)$ (does/does not) represent a member of the set of integers, because there is an integer q, namely 0, such that $b \cdot q = 0$. Hence, $\dfrac{0}{b} =$ ____ .

---

does; 0

Remark. This concludes the study of the system of integers. Try the self-evaluation test that follows. Score your paper from the answers given on page 278.

UNIT III
Self-Evaluation Test

1. The set of all natural numbers and their negatives, together with zero, is called the set of _____ .
2. If J = {integers} and W = {whole numbers}, then _____ ⊂ _____ .
3. The set of integers is a(n) (finite/infinite) _____ set.
4. For each integer a, $-a$ and a are called _____ _____ of each other.
5. If a is an integer, $-a$ (always/sometimes/never) denotes a negative integer.
6. The relative order of the integers -5, 7, and -9 can be shown by writing _____ < _____ < _____ .
7. The axioms (and theorems) for the system of whole numbers are (always/sometimes/never) applicable to the system of integers.
8. Which of the following axioms is a part of the system of integers, and is not in the system of whole numbers?
    a. the Commutative law of addition
    b. the Closure law for multiplication
    c. the Additive inverse law
9. On the number line

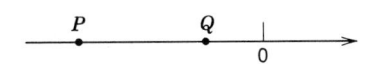

    the integer associated with point P is (less than/greater than) the integer associated with point Q.
10. "Natural numbers" and "positive integers" (always/sometimes/never) mean the same thing.
11. If a is an integer, the negative of a and the additive inverse of a (always/sometimes/never) represent the same number.
12. The sum of two negative integers is (always/sometimes/never) a negative integer.

144

The Set of Integers

13. |−5| is read "the _____ _____ of negative five."
14. The sum of a positive integer and a negative integer is (always/sometimes/never) a negative integer.
15. The sum of any integer a and zero is (zero/undefined/a).
16. Write $-4 - 2 + 8$ as a basic numeral.
17. The product of a positive integer and a negative integer is (always/sometimes/never) a positive integer.
18. The product of two negative integers is (always/sometimes/never) a positive integer.
19. Write $(-4) \cdot (-2) \cdot (7)$ as a basic numeral.
20. The product of any integer, a, and 0 is (zero/undefined/a).
21. If a and b are integers, $a - b$ is (always/sometimes/never) equal to $a + (-b)$.
22. If a and b are integers, $a - b$ is (always/sometimes/never) an integer.
23. If a and b are integers ($b \neq 0$), $\frac{a}{b}$ is (always/sometimes/never) an integer.
24. $\frac{0}{8} =$ _____ .
25. $\frac{8}{0}$ is _____ .

If you missed fewer than seven questions on this test, you are ready to continue on to Unit IV of this program. If you missed seven or more questions, you would probably profit by returning to the remark on page 107 and reading through the program to this point.

UNIT IV
The Set of Rational Numbers

OBJECTIVES

Upon completion of this part of the program, the reader will be able to:

1. Use the correct vocabulary and/or the correct notation relating to the following: rational number, multiplicative inverse or reciprocal, fraction, numerator, and denominator.

2. Identify the axiom adopted for rational numbers that characterizes the members in the set.

3. Associate rational numbers with points on a number line.

4. Determine whether two fractions represent the same number.

5. Identify the appropriate property of the set of rational numbers that justifies writing a fraction in an equivalent form.

6. Reduce fractions to lower terms and raise fractions to higher terms.

7. Identify the property of the set of rational numbers that justifies writing a sum, product, difference, or quotient in an equivalent form.

8. Express sums of rational numbers and products of rational numbers as single fractions.

9. Illustrate the product of two rational numbers using a lattice.

10. Express the difference of two rational numbers and the quotient of two rational numbers as single fractions.

Remark. We began our study of number systems by considering the set of whole numbers. We extended this set by inserting the negatives of the natural numbers to form the set of integers. We adopted for the system of integers the set of axioms we had employed in the system of whole numbers. To this we added an axiom governing the behavior of the negatives of the natural numbers with respect to the operation of addition.

One result of extending the set of whole numbers to the set of integers was that the new set $J = \{\ldots -3, -2, -1, 0, 1, 2, 3, \ldots\}$ was closed for subtraction. The set, however, was not closed for division.

Now let us extend the set of integers by inserting some new numbers. This extended set will of course, still contain integers as members, but, in addition, will contain such new numbers as are necessary to insure that *every pair of integers will have a quotient*. That is, $\frac{a}{b}$, where a and b are integers (except that b cannot be 0) will represent one of our new numbers, and conversely, every one of our new numbers can be viewed as the quotient of two integers. This extended set of numbers, called the **set of rational numbers**, contains the integers and all other quotients $\frac{a}{b}$, where $a, b \in J$ ($b \neq 0$).

First, we shall take as an axiom the following:

For all $b \in J$ ($b \neq 0$), there exists exactly one number $\frac{1}{b}$, such that

$$b \cdot \frac{1}{b} = 1 \text{ and } \frac{1}{b} \cdot b = 1.$$

This axiom is called the **Multiplicative inverse law**. It asserts, for example, that there is a number $\frac{1}{5}$ such that $5 \cdot \frac{1}{5} = 1$.

1. Similarly, there is a number $\frac{1}{4}$ such that $4 \cdot \frac{1}{4} =$ ____.

1

2. If the product of two numbers equals 1, then each of the factors in the product is said to be the **multiplicative inverse** of the other. By the _____ inverse law, $3 \cdot \frac{1}{3} = 1$. Hence, $\frac{1}{3}$ is the _____ _____ of 3.

Multiplicative; multiplicative inverse

148 UNIT IV

3. The word **reciprocal** is used in the same sense as the words "multiplicative inverse." Because $9 \cdot \frac{1}{9} = 1$, the number 9 is the reciprocal of $\frac{1}{9}$ and $\frac{1}{9}$ is the _____ of 9.

---

reciprocal

4. The multiplicative inverse, or reciprocal, of 7 is ____.

$\frac{1}{7}$

5. The multiplicative inverse of $\frac{1}{-3}$ is ____.

-3

6. Because, for every nonzero integer b, there exists a number $\frac{1}{b}$, then, every integer, except 0, has a multiplicative _____ or _____.

---

inverse; reciprocal

7. $b \cdot \frac{1}{b} = $ ____ $(b \neq 0)$.

1

Remark. Since we want to include the integers in the new set whose members include the reciprocals of all integers (except 0), we want the new set to have the properties (axioms and theorems) pertaining to the set of integers which are listed in Panels I and II on pages 81 and 135. These axioms and theorems lead to a new interpretation of the quotient $a \div b$ or $\frac{a}{b}$ where a and b are integers.

8. Recall that in our definition of a quotient in the set of integers, $6 \div 2$, or $\frac{6}{2}$, is the number q such that $2 \cdot q = $ ____.

$2 \cdot q = 6$

The Set of Rational Numbers 149

9. Consistent with this definition, the quotient
$$2 \div 5 = q \text{ or } \frac{2}{5} = q,$$
is the number such that $5 \cdot q = $ _____.

$5 \cdot q = 2$

10. Multiplying both members of the equality $5q = 2$ by $\frac{1}{5}$ gives
$$\frac{1}{5}(5q) = \frac{1}{5} \cdot \underline{}.$$

$\frac{1}{5} \cdot 2$

11. Application of the Associative and Commutative laws of multiplication yields
$$\left(\frac{1}{5} \cdot \underline{}\right) \cdot q = 2 \cdot \frac{1}{5}.$$

$\left(\frac{1}{5} \cdot 5\right) \cdot q$

12. By the Multiplicative inverse law, $\frac{1}{5} \cdot 5 = 1$. Substituting 1 for $\frac{1}{5} \cdot 5$ in the left-hand member of $\left(\frac{1}{5} \cdot 5\right) \cdot q = 2 \cdot \frac{1}{5}$ gives $1 \cdot q = $ _____.

$1 \cdot q = 2 \cdot \frac{1}{5}$

13. Substituting q for $1 \cdot q$ in $1 \cdot q = 2 \cdot \frac{1}{5}$ yields
$$q = \underline{}.$$

$q = 2 \cdot \frac{1}{5}$

14. Substituting $\frac{2}{5}$ for q from Frame 9 in the equation in $q = 2 \cdot \frac{1}{5}$ gives
$$\frac{2}{5} = \underline{}.$$

$\frac{2}{5} = 2 \cdot \frac{1}{5}$

UNIT IV

Remark. The foregoing sequence of frames suggests a theorem which gives an equivalent expression for the quotient $a \div b$ or $\frac{a}{b}$.

Theorem: If $a, b \in J$ $(b \neq 0)$, then $\frac{a}{b} = a \cdot \frac{1}{b}$.

Proof:

Let $\frac{a}{b} = q$.

$b \cdot q = a$ (Definition of a quotient)

$\frac{1}{b} \cdot (b \cdot q) = \frac{1}{b} \cdot a$ (Theorem: If $a = b$, then $c \cdot a = c \cdot b$)

$\left(\frac{1}{b} \cdot b\right) \cdot q = a \cdot \frac{1}{b}$ (Associative and Commutative laws of multiplication)

$1 \cdot q = a \cdot \frac{1}{b}$ (Multiplicative inverse law)

$q = a \cdot \frac{1}{b}$ (Identity law of multiplication)

$\frac{a}{b} = a \cdot \frac{1}{b}$.   (Substitution of $\frac{a}{b}$ for q)

We can consider a quotient $\frac{a}{b}$ ($b \neq 0$) to be equal to the product $a \cdot \frac{1}{b}$. From the proof above we are assured that this result is consistent with the definitions of a quotient we made in the set of whole numbers and in the set of integers.

15. Since $\frac{a}{b} = a \cdot \frac{1}{b}$ for every integer a and every nonzero integer b, we can write $3 \div 5$ or $\frac{3}{5}$ as $3 \cdot \frac{1}{5}$. Similarly, $\frac{4}{5} = $ _____ $\cdot$ _____. Furthermore, $4 \div 5 = $ _____.

$4 \cdot \frac{1}{5}$; $4 \cdot \frac{1}{5}$

16. $\frac{7}{9} = 7 \cdot$ _____ ; also $7 \div 9 = $ _____.

$7 \cdot \frac{1}{9}$; $7 \cdot \frac{1}{9}$

The Set of Rational Numbers   151

17. $\dfrac{4}{7} = 4 \cdot$ _____ ; also $4 \div 7 =$ _____ .

$4 \cdot \dfrac{1}{7}$; $4 \cdot \dfrac{1}{7}$

Remark. We have now extended the set of integers by including every quotient $\dfrac{a}{b}$, where a and b are integers and $b \neq 0$. This new set of numbers, called the set of **rational numbers**, we shall denote by Q. Hence

$$Q = \{\text{rational numbers}\} = \{\tfrac{a}{b} \mid a,b \in J, b \neq 0\}.$$

18. $\dfrac{2}{3}, \dfrac{1}{7},$ and $-\dfrac{15}{4}$ are examples of _____ numbers.

---

rational

Remark. We have assumed that each nonzero *integer* has a multiplicative inverse. Now let us extend the **Multiplicative inverse law** to all rational numbers. We shall assume that:

For all $\dfrac{a}{b} \in Q \left(\dfrac{a}{b} \neq 0\right)$, there is exactly one number $\dfrac{1}{\frac{a}{b}}$ with the property that

$$\dfrac{a}{b} \cdot \dfrac{1}{\frac{a}{b}} = 1.$$

19. From the Multiplicative inverse axiom, $\dfrac{3}{4}$ has the multiplicative inverse $\dfrac{1}{\frac{3}{4}}$, such that $\dfrac{3}{4} \cdot \dfrac{1}{\frac{3}{4}} =$ _____ .

1

20. By the Multiplicative inverse law, $\dfrac{2}{3} \cdot \dfrac{1}{\text{___}} = 1.$

$\dfrac{2}{3} \cdot \dfrac{1}{\frac{2}{3}} = 1$

21. The number $\frac{2}{3}$ is a rational number because 2 and 3 are _____.

---
integers

22. A rational number such as $\frac{2}{3}$ is an abstraction, just as the whole number 4 is an abstraction. The symbol used to denote such a number is called a **fraction**. Hence, the symbol "$\frac{2}{3}$" is a _____.

---
fraction

Remark. Note again that there is a distinction between a number and its name (symbol). As we did earlier, we might use quotation marks wherever we are referring to the symbol. We have elected instead to rely on the context without the use of quotation marks.

Now, before you get the impression that all fractions represent rational numbers, let us hasten to observe that such is not the case. Every rational number can be represented by a fraction, but there are many fractions that do not represent rational numbers. For example, consider the well-known symbol π (pi). Any fraction denoting quotients such as $\frac{\pi}{2}, \frac{4}{\pi}$, and $\frac{\pi}{-3}$ do not represent rational numbers. This is because the number π is not an integer.

In our discussions, all fractions will be presumed to represent rational numbers; that is, whenever and wherever we use variables such as a, b, c, or d hereafter, these symbols will denote integers, and integers only.

23. In a fraction $\frac{a}{b}$, a is called the **numerator** of the fraction, and b is called the **denominator** of the fraction. Thus, in the fraction $\frac{7}{9}$, 7 is the _____ and 9 is the _____.

---
numerator; denominator

24. Rational numbers are quotients of integers. That is, $\frac{a}{b}$ ($b \neq 0$), is the quotient of the integer ____ divided by the nonzero integer ____. The numerator of the fraction is ____ and the denominator is ____.

---
a; b; a; b

The Set of Rational Numbers   153

Remark. As we extended the set of integers, they have become members in good standing of the new set of rational numbers, as indeed do the whole numbers. Let us see what this implies.

25. In the set of rational numbers, Q, $\frac{4}{1}$ and 4 denote the same number. The symbol $\frac{5}{1}$ and _____ denote the same number.

5

26. Because the integers 7 and -9 are also quotients of integers $\frac{7}{1}$ and $\frac{-9}{1}$, they are also _____ numbers.

---
rational

27. A symbol such as 7 represents a natural number, an _____, and a _____ number.

---
integer; rational

28. Every integer (is/is not) a rational number.

---
is

29. 0 is an integer, therefore 0 (is/is not) a rational number.

---
is

30. The quotient $\frac{0}{b}$ ($b \neq 0$) can be denoted by the basic numeral _____, and is a rational number. However, the symbol $\frac{a}{0}$ is _____ for all rational numbers a.

0; undefined

Remark. It should be evident to you that there are infinitely many rational numbers. Indeed, since $\frac{1}{b}$ denotes a rational number for every nonzero integer b, and since the

▼

set of nonzero integers is infinite, there are infinitely many rational numbers that can be represented in this form alone.

When we discussed infinite sets earlier, you were able to employ the symbol ... to indicate that the set contained an infinite number of members. For example, we used $\{0, 1, 2, 3, \ldots\}$ to represent the infinite set of whole numbers. We do not, however, use this notation to denote the set of rational numbers, because there is no convenient way we can list a few of the members to establish a sequential pattern. Therefore, we simply use the notation $\{\text{rational numbers}\}$ or the set-builder notation $\{\frac{a}{b} \mid a, b \in J, b \neq 0\}$ to refer to the infinite set of rational numbers.

31. Each natural number is a member of $\{1, 2, 3, \ldots\}$, as well as a member of $\{\text{rational numbers}\}$. The set of natural numbers is, therefore, a subset of the set of _____ numbers. Using the subset symbol, we have that $\{1, 2, 3, \ldots\}$ ___ $\{\text{rational numbers}\}$, or N ___ Q.

---

rational; $\subset$; $\subset$

32. The set of integers is also a subset of the set of rational numbers. Represent this relationship using a subset symbol.

$\{\ldots -2, -1, 0, 1, 2, \ldots\} \subset \{\text{rational numbers}\}$
(Or $\{\text{integers}\} \subset \{\text{rational numbers}\}$, or $J \subset Q$)

Remark. The next sequence of frames will give you an opportunity to review some fundamental ideas about rational numbers.

R33. The axiom: For all $b \in J$ $(b \neq 0)$, $b \cdot \frac{1}{b} = 1$, is called the Multiplicative _____ law.

---

inverse

R34. The number $\frac{1}{5}$ is the _____ of 5, because $5 \cdot \frac{1}{5} = 1$.

---

multiplicative inverse

R35. If $a, b \in J$ $(b \neq 0)$, then $\frac{a}{b} = a \cdot$ _____

$\frac{a}{b} = a \cdot \frac{1}{b}$

The Set of Rational Numbers   155

R36. The quotient $\frac{2}{7}$ can be written as the product $2 \cdot$ _____. Similarly, $\frac{5}{9} =$ _____.

$2 \cdot \frac{1}{7}$; $5 \cdot \frac{1}{9}$

R37. If a and b are integers and if $b \neq 0$, then the symbol $\frac{a}{b}$ denotes the quotient of a and b. Every such quotient is a _____ number.

---
rational

R38. The set of rational numbers can be denoted by $Q = \{$ _____ $\}$.

{rational numbers}   (Or $\{\frac{a}{b} \mid a, b \in J, b \neq 0\}$.)

R39. The set of rational numbers is a(n) (finite/infinite) set.

---
infinite

R40. The set of integers (is/is not) a subset of the set of rational numbers.

---
is

R41. For every nonzero rational number $\frac{a}{b}$, there is exactly one number $\frac{1}{\frac{a}{b}}$ called the _____ _____ of $\frac{a}{b}$.

---
multiplicative inverse   (Or reciprocal.)

R42. The axioms for the rational number system differ from the axioms adopted for the system of integers only in the inclusion of the _____ _____ axiom, which is not included in the axioms for the integers.

Multiplicative inverse

R43. A numeral such as $\frac{"6"}{11}$ used to denote the rational number $\frac{6}{11}$ is called a _____.
In this example, 6 is the _____ and 11 is the _____.

---
fraction; numerator; denominator

UNIT IV

R44. If b is any nonzero integer, $\dfrac{b}{0}$ is _____, and $\dfrac{0}{b} =$ _____.

---

undefined; 0

Remark. The next sequence of frames will give you an opportunity to review some of the axioms which were adopted for the set of integers and which we now adopt for the set of rational numbers. You may want to refer to Panels I and II on pages 81 and 135 as you proceed through the remainder of this part of the program. In the following frames, assume that all fractions represent rational numbers.

R45. The Commutative law of addition states that $\dfrac{a}{b} + \dfrac{c}{d} = \dfrac{c}{d} +$ _____. For example, $\dfrac{2}{3} + \dfrac{3}{4} = \dfrac{3}{4} +$ _____.

$\dfrac{a}{b}$; $\dfrac{2}{3}$

R46. The Associative law of addition states that $\left(\dfrac{a}{b} + \dfrac{c}{d}\right) + \dfrac{e}{f} = \dfrac{a}{b} + \left(\underline{}\right)$.

For example, $\left(\dfrac{7}{9} + \dfrac{3}{5}\right) + \dfrac{1}{12} = \dfrac{7}{9} + \left(\underline{}\right)$.

$\left(\dfrac{c}{d} + \dfrac{e}{f}\right)$; $\left(\dfrac{3}{5} + \dfrac{1}{12}\right)$

R47. The Commutative law of multiplication states that $\dfrac{a}{b} \cdot \dfrac{c}{d} = \dfrac{c}{d} \cdot$ _____.

For example, $\dfrac{2}{7} \cdot \dfrac{5}{13} = \dfrac{5}{13} \cdot$ _____.

$\dfrac{a}{b}$; $\dfrac{2}{7}$

R48. The Associative law of multiplication states that $\left(\dfrac{a}{b} \cdot \dfrac{c}{d}\right) \cdot \dfrac{e}{f} = \dfrac{a}{b} \cdot \left(\underline{}\right)$.

For example, $\left(\dfrac{4}{7} \cdot \dfrac{1}{5}\right) \cdot \dfrac{3}{8} = \dfrac{4}{7} \cdot \left(\underline{}\right)$.

$\dfrac{a}{b} \cdot \left(\dfrac{c}{d} \cdot \dfrac{e}{f}\right)$; $\left(\dfrac{1}{5} \cdot \dfrac{3}{8}\right)$

The Set of Rational Numbers

R49. The Distributive law states that $\frac{e}{f} \cdot \left(\frac{a}{b} + \frac{c}{d}\right) = \frac{e}{f} \cdot \frac{a}{b} + \underline{}$.

For example, $\frac{2}{9} \cdot \left(\frac{3}{11} + \frac{7}{8}\right) = \frac{2}{9} \cdot \frac{3}{11} + \underline{}$.

$\frac{e}{f} \cdot \frac{c}{d}; \frac{2}{9} \cdot \frac{7}{8}$

R50. The Identity law for addition states that for every rational number $\frac{a}{b}$, $\frac{a}{b} + \underline{} = \frac{a}{b}$.

For example, $\frac{5}{6} + \underline{} = \frac{5}{6}$.

0; 0

R51. The Identity law for multiplication states that for every rational number $\frac{a}{b}$, $\frac{a}{b} \cdot \underline{} = \frac{a}{b}$. For example, $\frac{7}{9} \cdot \underline{} = \frac{7}{9}$.

1; 1

R52. The Additive inverse law states that for every rational number $\frac{a}{b}$, $\frac{a}{b} + \left(-\frac{a}{b}\right) = \underline{}$.

For example, $\frac{1}{13} + \left(-\frac{1}{13}\right) = \underline{}$.

0; 0

R53. The Multiplicative inverse law states that for every nonzero rational number $\frac{a}{b}$, $\frac{a}{b} \cdot \frac{1}{\frac{a}{b}} = \underline{}$. For example, $\frac{6}{11} \cdot \frac{1}{\frac{6}{11}} = \underline{}$.

1; 1

Remark. See Exercise IVa, page 264, for additional practice on the topics considered to this point in this part of the program.

The remainder of this part of the program will be concerned with the very important and useful consequences of the axioms that we have adopted for the system of rational numbers. These consequences justify the routine symbol manipulations that characterize the study of rational numbers in the early school years. First let us see how every rational number can be represented by infinitely many fractions.

158 UNIT IV

54. Recall that $\frac{4}{4} = 1$, because 4 · 1 = 4. Similarly, $\frac{1}{1} = 1$ because 1 · 1 = 1, $\frac{2}{2} = 1$ because 2 · 1 = 2, and $\frac{3}{3} = 1$ because 3 · 1 = 3. Indeed, for every nonzero integer b, $\frac{b}{b} = 1$ because _____ · _____ = _____.

$b \cdot 1 = b$

55. We have an endless number of different fractions representing the number denoted by the basic numeral 1. The fraction $\frac{22}{22}$ is another name for the number whose basic numeral is _____.

1

56. Because 4 · 3 = 12, we can write $\frac{12}{4}$ = _____.

3

57. The fractions $\frac{12}{4}$, $\frac{18}{6}$, and $\frac{15}{5}$ are all names for the same number. The basic numeral for the number is _____.

3

Remark. The previous frames suggest that, at least with respect to a rational number that is also an integer, we can produce as many different names for the number as we wish. Before we extend this notion to apply to every rational number, let us turn our attention to the representation of rational numbers on the number line.

58. Consider the number line:

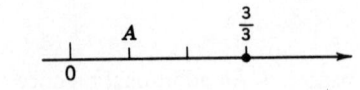

Recall that $\frac{3}{3} = 3 \cdot \frac{1}{3}$, and that we can view $3 \cdot \frac{1}{3}$ as denoting three one-thirds. This being the case, the end point of the first segment to the right of the origin, labeled A, would logically correspond to the rational number _____.

$\frac{1}{3}$

The Set of Rational Numbers   159

59. Bearing in mind that $\frac{2}{3}$ can be interpreted as two one-thirds, and that $\frac{0}{3}$ is another representation for 0, label each indicated point on the following number line with a fraction whose denominator is 3.

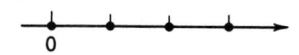

$\frac{0}{3} \quad \frac{1}{3} \quad \frac{2}{3} \quad \frac{3}{3}$

60. The point labeled B on the number line

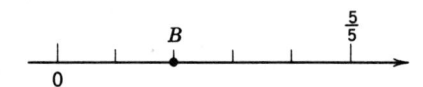

corresponds to the rational number _____.

$\frac{2}{5}$

61. The graph of $\frac{3}{5}$ is shown on the number line below. Show the graph of $\frac{8}{5}$.

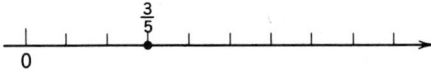

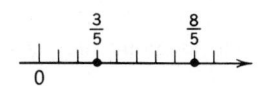

$\frac{3}{5} \quad \frac{8}{5}$

62. Use the number line below to show the graph of the rational number $\frac{8}{3}$.

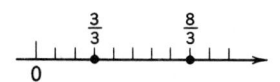

$\frac{3}{3} \quad \frac{8}{3}$

Remark. Observe that, in these few frames, we have been labeling the point on the number line corresponding to 1 with an appropriate fraction, e.g., $\frac{3}{3}$. Because, by the definition of a quotient, $\frac{6}{3} = 2$, we could also label the point corresponding to 2 with the fraction $\frac{6}{3}$. Moreover, in the event we were interested in finding graphs of rational numbers denoted by fractions with denominator 4, we could use $\frac{4}{4}$ for 1, $\frac{8}{4}$ for 2, and so on. Thus, we can use $\frac{1}{1}, \frac{2}{2}, \frac{3}{3}$, etc. to denote 1; $\frac{2}{1}, \frac{4}{2}, \frac{6}{3}$, etc. to denote 2; and, in fact, we can represent any of the integers in the rational numbers by any number of fractions we please. Therefore, we can label the corresponding points on a number line as follows:

and can continue supplying such names for as long as we wish. Now we want to argue that this also applies to the point corresponding to any rational number.

63. Each of the segments *OA, AB, BC,* and *CD* on the number line

is composed of two of the duplicated segments used to locate points corresponding to $\frac{1}{8}, \frac{2}{8}, \frac{3}{8}$, etc. Because they each contain the same number of the original segments, *OA, AB, BC,* and *CD* are duplicates of each other. There are *four* of these larger duplicated segments lying between the points corresponding to 0 and 1. The end points of these segments, therefore, correspond to the rational numbers $\frac{1}{4}$, ___ , ___ , and $\frac{4}{4}$.

$\frac{2}{4}, \frac{3}{4}$

The Set of Rational Numbers 161

64. The results of the preceding frame indicate, then, that we can label the appropriate points on a number line as follows:

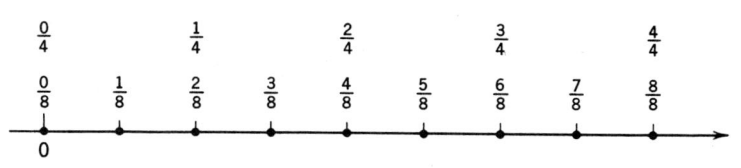

[number line showing $\frac{0}{4}, \frac{1}{4}, \frac{2}{4}, \frac{3}{4}, \frac{4}{4}$ aligned with $\frac{0}{8}, \frac{1}{8}, \frac{2}{8}, \frac{3}{8}, \frac{4}{8}, \frac{5}{8}, \frac{6}{8}, \frac{7}{8}, \frac{8}{8}$ above 0]

Since $\frac{1}{4}$ and $\frac{2}{8}$ are different names for the same number, we can write $\frac{1}{4} = \frac{2}{8}$.

Similarly, $\frac{2}{4} = $ ____ and $\frac{3}{4} = $ ____ .

$\frac{4}{8}$; $\frac{6}{8}$

65. Consider the following number line, which is constructed according to the process used in the preceding frame:

[number line showing fractions with denominators 8, 4, 2 above integers 0, 1, 2, 3, with point A marked]

The point labeled A on this number line can be labeled with any of the fractions

____ , ____ , or ____ .

$\frac{1}{2}$; $\frac{2}{4}$; $\frac{4}{8}$ (*Any order is correct.*)

66. From the preceding frame, it is apparent that $\frac{1}{2}, \frac{2}{4},$ and $\frac{4}{8}$ are all fractions denoting the same _____ number.

---

rational

Remark. Although it is possible to arrange the integers in a discrete arrangement, ... $-3, -2, -1, 0, 1, 2, 3, ...$, each number having a distinct neighbor to the left and

▼

162 UNIT IV

to the right, this is not the case with rational numbers. The previous sequence of frames suggests that between every pair of rational numbers there is another rational number. There are an infinite number of members between any two members. This property is described by saying that the set of rational numbers is *dense*.

As we observed in the previous frames, the number line also makes it plausible that we can find infinitely many different fractions denoting the *same* rational number. We now wish to explore a more general way of determining when two fractions are equal, that is, when two fractions represent the same rational number.

While the number line is an aid to our intuition, it will never get us beyond a few specific cases, and we would like to have complete generality. As usual, this can be obtained only by an appeal to the axioms for rational numbers.

67. From the number line we observe that $\frac{1}{2} = \frac{2}{4}$, $\frac{1}{2} = \frac{4}{8}$, and $\frac{2}{4} = \frac{4}{8}$. A careful study of these statements of equality will show that, in each one of them, the products formed by using as factors the numerator of one fraction and the denominator of the other are equal. That is,

$$\frac{1}{2} = \frac{2}{4}, \text{ and } 1 \cdot 4 = 2 \cdot 2;$$

$$\frac{1}{2} = \frac{4}{8}, \text{ and } 1 \cdot 8 = 4 \cdot 2;$$

$$\frac{2}{4} = \frac{4}{8}, \text{ and } 2 \cdot 8 = \underline{\qquad} \cdot \underline{\qquad}.$$

$2 \cdot 8 = 4 \cdot 4$

68. The number line

$$\begin{array}{c} \frac{0}{3} \quad\quad \frac{1}{3} \quad\quad \frac{2}{3} \quad\quad \frac{3}{3} \\ \frac{0}{6} \;\; \frac{1}{6} \;\; \frac{2}{6} \;\; \frac{3}{6} \;\; \frac{4}{6} \;\; \frac{5}{6} \;\; \frac{6}{6} \\ 0 \end{array}$$

shows us that $\frac{1}{3} = \frac{2}{6}$, and $\frac{2}{3} = \underline{\qquad}$.

$\frac{4}{6}$

The Set of Rational Numbers   163

69. $\frac{1}{3} = \frac{2}{6}$, and $1 \cdot 6 = 2 \cdot 3$. Similarly,

$$\frac{2}{3} = \frac{4}{6}, \text{ and } \underline{\quad \cdot \quad} = \underline{\quad \cdot \quad}.$$

$2 \cdot 6 = 4 \cdot 3$

Remark. In general, it appears that, whenever one fraction, $\frac{a}{b}$, denotes the same rational number as another, $\frac{c}{d}$, then $a \cdot d = c \cdot b$. (Of course, as usual, b and d are not 0.) As it happens, this is indeed a valid statement, and can be shown to be a consequence of our axioms.

70. Let us assume that $\frac{2}{3} = \frac{4}{6}$ is true. If it is, then $2 \cdot \frac{1}{3} = 4 \cdot \frac{1}{6}$. Now, let us apply the theorem that permits us to multiply both members of an equality by the same rational number and multiply each member of

$$2 \cdot \frac{1}{3} = 4 \cdot \frac{1}{6}$$

by $(3 \cdot 6)$. The result is

$$(3 \cdot 6) \cdot \left(2 \cdot \frac{1}{3}\right) = (\underline{\quad \cdot \quad}) \cdot \left(4 \cdot \frac{1}{6}\right).$$

$(3 \cdot 6)$

71. By applying the Commutative and Associative laws of multiplication to each member of the above equation,

$$(3 \cdot 6) \cdot \left(2 \cdot \frac{1}{3}\right) = (3 \cdot 6) \cdot \left(4 \cdot \frac{1}{6}\right) \qquad (1)$$

we have

$$(2 \cdot 6) \cdot \left(3 \cdot \frac{1}{3}\right) = (4 \cdot 3) \cdot \left(6 \cdot \frac{1}{6}\right). \qquad (2)$$

Because $3 \cdot \frac{1}{3} = 1$ and $6 \cdot \frac{1}{6} = 1$, Equation (2) can be written

$$(2 \cdot 6) \cdot 1 = (4 \cdot 3) \cdot 1. \qquad (3)$$

▼

By the Multiplicative identity law, we can write Equation (3) as
$$2 \cdot 6 = \underline{} \cdot \underline{}.$$

$2 \cdot 6 = 4 \cdot 3$ *(This is what we wanted to show.)*

Remark. The axioms invoked in the last two frames can be applied in a similar way to the general case.

Theorem: If $a, b, c, d \in J$ $(b, d \neq 0)$, and if $\frac{a}{b} = \frac{c}{d}$, then $a \cdot d = c \cdot b$.

Proof:

$\dfrac{a}{b} = \dfrac{c}{d}$ (given information)

$a \cdot \dfrac{1}{b} = c \cdot \dfrac{1}{d}$ $\left(\text{Theorem: } \dfrac{a}{b} = a \cdot \dfrac{1}{b}\right)$

$(b \cdot d) \cdot \left(a \cdot \dfrac{1}{b}\right) = (b \cdot d) \cdot \left(c \cdot \dfrac{1}{d}\right)$ (Theorem: If $a = b$, then $c \cdot a = c \cdot b$.)

$(a \cdot d) \cdot \left(b \cdot \dfrac{1}{b}\right) = (c \cdot b) \cdot \left(d \cdot \dfrac{1}{d}\right)$ (Commutative and Associative laws of multiplication)

$(a \cdot d) \cdot 1 = (c \cdot b) \cdot 1$ (Multiplicative inverse axiom)

$a \cdot d = c \cdot b$ (Multiplicative identity axiom)

This gives us a criterion by which we can identify fractions that represent the same rational number. It is equally true that, if $a \cdot d = c \cdot b$, then $\frac{a}{b} = \frac{c}{d}$. This can be demonstrated simply by reversing the steps of the above proof. We state it formally as follows.

If $a, b, c, d \in J$ $(b, d \neq 0)$, and if $a \cdot d = c \cdot b$, then $\frac{a}{b} = \frac{c}{d}$.

Let us apply the theorem a few times for practice.

72. $\dfrac{3}{7} = \dfrac{6}{14}$, because $\underline{} \cdot \underline{} = \underline{} \cdot \underline{}$.

$3 \cdot 14 = 6 \cdot 7$

73. $\dfrac{8}{5}$ (is/is not) equal to $\dfrac{9}{6}$.

---

is not $(8 \cdot 6 \neq 9 \cdot 5)$

The Set of Rational Numbers 165

74. $\frac{4}{7}$ (is/is not) equal to $\frac{16}{28}$.

---

is $(4 \cdot 28 = 16 \cdot 7.)$

Remark. Now, we have an easy way of determining whether two fractions represent the same number. Our next step is to find ways of generating fractions representing the same number. First consider the following numerical example.

75. The Associative and Commutative laws of multiplication justify writing products such as $(2 \cdot 4) \cdot 3$ as ____ $\cdot (3 \cdot 4)$.

$2 \cdot (3 \cdot 4)$

76. The previous theorem asserts that if $a \cdot d = c \cdot b$, then $\frac{a}{b} = \frac{c}{d}$ ($b, d \neq 0$). Hence,

$$\text{if } (2 \cdot 4) \cdot 3 = 2 \cdot (3 \cdot 4), \text{ then } \frac{2 \cdot 4}{3 \cdot 4} = \frac{2}{3}.$$

Similarly,

$$\text{if } (2 \cdot 5) \cdot 3 = 2 \cdot (3 \cdot 5), \text{ then } \frac{2 \cdot 5}{3 \cdot 5} = \underline{\quad}.$$

$\frac{2 \cdot 5}{3 \cdot 5} = \frac{2}{3}$

Remark. The preceding frames suggest the following theorem:

Theorem: If $a, b, c \in J$ ($b, c \neq 0$), then $\frac{a \cdot c}{b \cdot c} = \frac{a}{b}$.

Proof:
$a \cdot (b \cdot c) = a \cdot (b \cdot c)$ (Reflexive law of equality)
$(a \cdot c) \cdot b = a \cdot (b \cdot c)$ (Commutative and Associative laws of multiplication)
$\frac{a \cdot c}{b \cdot c} = \frac{a}{b}$ Theorem: If $a \cdot d = c \cdot b$, then $\frac{a}{b} = \frac{c}{d}$.

77. For convenience, we shall refer to the theorem in the previous remark as the **Fundamental principle of fractions**. The Symmetric law of equality then

▼

permits writing the statement of the theorem in the form

$$\frac{a}{b} = \frac{a \cdot \_\_\_}{b \cdot \_\_\_}.$$

$\frac{a}{b} = \frac{a \cdot c}{b \cdot c}$

78. The statement $\frac{a}{b} = \frac{a \cdot c}{b \cdot c}$ can be interpreted to mean that if the numerator and denominator of a fraction are each multiplied by the same (nonzero) integer, the resulting fraction denotes the same rational number as the original fraction. Since $\frac{7 \cdot 3}{8 \cdot 3} = \frac{21}{24}$, the _____ principle of fractions assures us that $\frac{7}{8}$ and $\frac{21}{24}$ are different names for the same _____ number.

Fundamental; rational

79. $\frac{7 \cdot \_\_}{8 \cdot \_\_} = \frac{28}{32}$. Therefore, $\frac{7}{8}$ and $\frac{28}{32}$ are different fractions representing the same _____ _____.

$\frac{7 \cdot 4}{8 \cdot 4}$; rational number

80. $\frac{11 \cdot 6}{5 \cdot 6} = \frac{66}{30}$. Therefore, $\frac{11}{5}$ and $\underline{}$ are different fractions representing the same rational number.

$\frac{66}{30}$

81. The Fundamental principle of fractions, $\frac{a}{b} = \frac{a \cdot c}{b \cdot c}$, when viewed as $\frac{a \cdot c}{b \cdot c} = \frac{a}{b}$, can be used to rewrite $\frac{8}{12}$ as $\frac{2}{3}$. That is, $\frac{8}{12} = \frac{2 \cdot 4}{3 \cdot 4} = \frac{2}{3}$. Similarly, $\frac{9}{15} = \frac{3 \cdot 3}{5 \cdot 3} = \underline{}$.

$\frac{3}{5}$

The Set of Rational Numbers

82. $\dfrac{8}{10} = \dfrac{4 \cdot 2}{5 \cdot 2} = \underline{}$.

$\dfrac{4}{5}$

Remark. Writing $\dfrac{8}{10} = \dfrac{4 \cdot 2}{5 \cdot 2} = \dfrac{4}{5}$ can be viewed as the division of 8 by 2 in the numerator of $\dfrac{8}{10}$ and 10 by 2 in the denominator to obtain $\dfrac{4}{5}$. That is

$$\dfrac{8}{10} = \dfrac{8 \div 2}{10 \div 2} = \dfrac{4}{5}.$$

In general, the fundamental principle of fractions can be stated as follows:

If the numerator and denominator of a fraction are each multiplied or divided by the same nonzero integer, the resultant fraction represents the same rational number as the original fraction.

This principle is most important, and is used very often in working with fractions.

83. The numerator and denominator of a fraction are frequently referred to as the **terms** of the fraction. $\dfrac{10}{15}$ is said to be in **higher terms** than $\dfrac{2}{3}$, because $\dfrac{10}{15}$ can be obtained from $\dfrac{2}{3}$ by multiplying 2 and 3 respectively by 5. Because $\dfrac{3 \cdot 7}{4 \cdot 7} = \dfrac{21}{28}$, $\dfrac{21}{28}$ is in \underline{} terms than $\dfrac{3}{4}$.

---

higher

84. If $a, b, c \in J$ ($b, c \neq 0$), $\dfrac{a \cdot c}{b \cdot c}$ can be obtained from $\dfrac{a}{b}$ by multiplying both a and b by c. Therefore, $\dfrac{a \cdot c}{b \cdot c}$ is in \underline{} than $\dfrac{a}{b}$.

---

higher terms

85. Because $\dfrac{a \cdot c}{b \cdot c}$ is in higher terms than $\dfrac{a}{b}$, then, conversely, $\dfrac{a}{b}$ is said to be in **lower terms** than $\dfrac{a \cdot c}{b \cdot c}$. Thus, $\dfrac{1}{2}$ is in lower terms than $\dfrac{2}{4}$ and $\dfrac{1}{3}$ is in (lower/higher) terms than $\dfrac{2}{6}$.

---

lower

86. The fraction $\frac{1}{5}$ is in _____ _____ than $\frac{2}{10}$.

---

lower terms

87. $\frac{a \cdot c}{b \cdot c} = \frac{a}{b}$ $(b, c \neq 0)$ is a statement of the _____ _____ of fractions, and the right-hand member, $\frac{a}{b}$, is in _____ terms than the left-hand member, $\frac{a \cdot c}{b \cdot c}$.

Fundamental principle; lower

88. A fraction with integers for numerator and denominator is said to be in **lowest terms** when the numerator and denominator contain no integral factors (other than 1 and −1) in common. Thus, $\frac{3}{4}$ is in lowest terms, while $\frac{9}{12}$ is not, because both 9 and 12 contain the common factor _____.

3

89. When a fraction is written in lower terms, the original fraction is said to have been **reduced**. For example, we can write $\frac{5}{30} = \frac{1 \cdot 5}{6 \cdot 5} = $ _____ . Alternatively, we can write $\frac{5}{30} = \frac{5 \div 5}{30 \div 5} = $ _____ .

$\frac{1}{6}$; $\frac{1}{6}$

90. At times, it is convenient to represent the division of the numerator and denominator by a common factor by use of a slant bar, /. Thus,

$$\frac{4}{6} = \frac{\cancel{4}^{2}}{\cancel{6}_{3}} = \frac{2}{3},$$

where the slant bar is used to indicate that 4 and 6 have each been _____ by the common factor 2.

---

divided

The Set of Rational Numbers   169

91. $\dfrac{24}{26} = \dfrac{\overset{12}{\cancel{24}}}{\underset{13}{\cancel{26}}} = \dfrac{12}{13}$, where the slant bars indicate that 24 and 26 have each been divided by ____ .

2

92. To reduce $\dfrac{13}{26}$ to lowest terms, both numerator and denominator would be _____ by ____ .

---

divided; 13

93. Observe the following forms:

$$\dfrac{6}{8} = \dfrac{3 \cdot 2}{4 \cdot 2} = \dfrac{3}{4}, \quad \dfrac{6}{8} = \dfrac{6 \div 2}{8 \div 2} = \dfrac{3}{4}, \quad \text{and} \quad \dfrac{6}{8} = \dfrac{\overset{3}{\cancel{6}}}{\underset{4}{\cancel{8}}} = \dfrac{3}{4}.$$

Each of these represents the reduction of $\dfrac{6}{8}$ to ____ .

$\dfrac{3}{4}$

Remark. Any of these forms may at one time or another be more convenient to use than the others. Because the first form parallels the statement of the Fundamental principle of fractions, this is the form we shall generally use.

94. $\dfrac{1}{2}$ is equal to $\dfrac{24}{48}, \dfrac{12}{24}, \dfrac{6}{12}$, and $\dfrac{3}{6}$. These fractions represent the same _____ number.

---

rational

95. $\dfrac{1}{2}$ is in _____ terms because the numerator and denominator contain no integral factor in common other than 1 and −1.

---

lowest

170   UNIT IV

96. Any fraction expressed in lowest terms is conveniently called a **basic fraction**. $\frac{3}{4}, \frac{6}{8}, \frac{9}{12}$, and $\frac{12}{16}$ represent the same rational number. Of these, ____ is the basic fraction.

$\frac{3}{4}$

97. Since $\frac{2}{3}$ is in lowest terms, it is a ____ fraction.

---

basic

Remark. The term "basic fraction," of course, simply means "basic numeral," because fractions are numerals. We shall use the term "basic fraction" on the grounds that it is a better aid to intuition when speaking of fractions. This means, however, that we want to call $\frac{2}{1}$, rather than 2, the basic fraction for $\frac{4}{2}$.

Let us go through a brief sequence of frames here to review some of the consequences of the axioms for the rational numbers.

R98. $\frac{a}{b}$ equals the product $a \cdot$ ____ .

$a \cdot \frac{1}{b}$

R99. The rational numbers can be associated with points on a line. On the number line

the number associated with the point labeled Q is $\frac{2}{3}$ and the number associated with the point labeled R is ____ .

$\frac{7}{3}$

The Set of Rational Numbers 171

R100. A basic numeral such as 4 (does/does not) represent a rational number.

does $\left(4 = \frac{4}{1}.\right)$

R101. $\frac{a}{b} = \frac{c}{d}$ $(b, d \neq 0)$ if _____ = _____ .

$a \cdot d = c \cdot b$ (Or $c \cdot b = a \cdot d$.)

R102. $\frac{4}{27}$ (does/does not) equal $\frac{35}{243}$.

does not $(4 \cdot 243 \neq 35 \cdot 27.)$

R103. For any nonzero number c, $\frac{a}{b} = \frac{a \cdot c}{b \cdot c}$. This theorem is called the _____ _____ of fractions.

Fundamental principle

R104. If one fraction can be obtained from another by applying the Fundamental principle of fractions in either of the forms

$$\frac{a}{b} = \frac{a \cdot c}{b \cdot c} \quad \text{or} \quad \frac{a \cdot c}{b \cdot c} = \frac{a}{b},$$

then the fractions (always/sometimes/never) denote the same rational number.

always

R105. The fraction $\frac{a \cdot c}{b \cdot c}$ is said to be in _____ terms than $\frac{a}{b}$, and $\frac{a}{b}$ is said to be in _____ terms than $\frac{a \cdot c}{b \cdot c}$.

higher; lower

R106. Each example of rewriting a given fraction in lower or higher terms represents an application of the _____ _____ of fractions.

Fundamental principle

R107. A fraction written in lowest terms is called a _____ fraction.

---

basic

R108. $\dfrac{12}{30} = \dfrac{2 \cdot 6}{5 \cdot 6} = \dfrac{2}{5}$; $\dfrac{12}{30} = \dfrac{12 \div 6}{30 \div 6} = \dfrac{2}{5}$; and $\dfrac{12}{30} = \dfrac{\overset{2}{\cancel{12}}}{\underset{5}{\cancel{30}}} = \dfrac{2}{5}$

are different representations of an application of the _____
_____ of fractions.

Fundamental principle

Remark. See Exercise IVb, page 267, for additional practice using the Fundamental principle of fractions.

Rational numbers have rarely been mentioned in arithmetic classes. All of the student's attention has been focused on the manipulation of the symbols (fractions) that denote these rational numbers. Thus, we frequently hear of "adding fractions," or "multiplying fractions" but not of sums or products of rational numbers, which is really what we are concerned with. However, it is important that you be able to perform routine manipulations with fractions (numerals) skillfully and with reasonable rapidity if you are to be able to use rational numbers in any practical way. The problem lies in the fact that if such manipulations are performed in a rote and meaningless way, the very concepts that have to be used to apply fractions to practical situations are completely overlooked. The structure of the rational number system, that is, the patterns it shares with other number systems, is extremely important in making work with rational numbers meaningful.

We first consider the addition of fractions. We shall rewrite sums of rational numbers in various ways to obtain representations in the form of basic fractions. Our point of view is that symbols such as $\dfrac{1}{5} + \dfrac{2}{5}$ already represent a sum, and writing the sum as $\dfrac{3}{5}$ (as we will show) simply represents the same number by means of a basic fraction.

Recall that the distributive law for rational numbers asserts that

$$\frac{e}{f} \cdot \left(\frac{a}{b} + \frac{c}{d}\right) = \frac{e}{f} \cdot \frac{a}{b} + \frac{e}{f} \cdot \frac{c}{d},$$

or, in the form in which we shall want it,

$$\frac{a}{b} \cdot \frac{e}{f} + \frac{c}{d} \cdot \frac{e}{f} = \left(\frac{a}{b} + \frac{c}{d}\right) \cdot \frac{e}{f}.$$

The Set of Rational Numbers 173

109. The sum of the rational numbers $\frac{2}{7}$ and $\frac{3}{7}$ equals $\frac{2}{7} + \frac{3}{7}$. The sum of $\frac{2}{3}$ and $\frac{5}{3}$ equals $\frac{2}{3} +$ _____ .

$\frac{2}{3} + \frac{5}{3}$

110. The sum $\frac{2}{3} + \frac{5}{3}$ can be written $2 \cdot \frac{1}{3} +$ _____ $\cdot \frac{1}{3}$.

5

111. The Distributive law justifies writing $2 \cdot \frac{1}{3} + 5 \cdot \frac{1}{3} = ($ _____ $+$ _____ $) \cdot \frac{1}{3}$.

$(2 + 5) \cdot \frac{1}{3}$

112. The sum $2 \cdot \frac{1}{3} + 5 \cdot \frac{1}{3} = (2 + 5) \cdot \frac{1}{3} =$ _____ $\cdot \frac{1}{3} =$ _____ .

$7 \cdot \frac{1}{3} = \frac{7}{3}$

113. The sum $\frac{4}{9} + \frac{3}{9} = 4 \cdot \frac{1}{9} + 3 \cdot \frac{1}{9}$. The _____ law justifies writing the right-hand member as $(4 + 3) \cdot \frac{1}{9}$ which can then be written as $\frac{4 + 3}{9}$ or $\frac{7}{9}$.

Distributive

Remark. The last few frames suggest the following theorem.

Theorem: If $a, b, c \in J$ $(c \neq 0)$, then
$$\frac{a}{c} + \frac{b}{c} = \frac{a + b}{c}.$$

Proof:
$$\frac{a}{c} + \frac{b}{c} = a \cdot \frac{1}{c} + b \cdot \frac{1}{c} \qquad \left(\text{Theorem: } \frac{a}{b} = a \cdot \frac{1}{b}\right)$$
$$= (a + b) \cdot \frac{1}{c} \qquad \text{(Distributive law)}$$
$$= \frac{a + b}{c} \qquad \left(\text{Theorem: } \frac{a}{b} = a \cdot \frac{1}{b}\right)$$

174 UNIT IV

114. The theorem in the preceding remark justifies writing the statement $\frac{2}{7} + \frac{3}{7} = \frac{2+3}{7}$.

Similarly $\frac{3}{5} + \frac{1}{5} = \frac{+}{}$.

$\frac{3+1}{5}$

115. The basic fractions for $\frac{2+3}{7}$ and $\frac{3+1}{5}$ in Frame 114 are ___ and ___, respectively.

$\frac{5}{7}; \frac{4}{5}$

116. The sum $\frac{3}{11} + \frac{2}{11} = \frac{3+2}{11}$, which can be written as the basic fraction $\frac{5}{11}$. The sum $\frac{4}{15} + \frac{7}{15} = \frac{+}{15}$, which can be written as the basic fraction ___.

$\frac{4}{15} + \frac{7}{15} = \frac{4+7}{15}; \frac{11}{15}$

117. It is customary, and usually more convenient, to represent the sums of rational numbers by basic fractions, that is, fractions in lowest terms. Thus, while $\frac{3}{8} + \frac{1}{8}$ can be rewritten as $\frac{3+1}{8} = \frac{4}{8}$, the basic fraction for the sum is ___.

$\frac{1}{2}$

118. $\frac{1}{6} + \frac{2}{6} = \frac{1+2}{6} = \frac{3}{6}$, which can be rewritten as the basic fraction ___.

$\frac{1}{2}$

Remark. To this point, we have concerned ourselves with the sums of rational numbers that involved fractions having the *same* denominators. We have been rewriting these sums by applying the theorem $\frac{a}{c} + \frac{b}{c} = \frac{a+b}{c}$. How can we rewrite the sum of two
▼

The Set of Rational Numbers 175

rational numbers such as $\frac{3}{8} + \frac{1}{4}$, where the fractions do not have the same denominator? We can first rewrite such sums as equal sums in which the fractions involved do have *like*, or *common*, denominators.

119. To find a common denominator for two or more fractions with **different**, or unlike, denominators, it is necessary to find an integer that is a multiple of each of the denominators. In other words, each denominator must be a factor of this common denominator. For example, a common denominator for $\frac{3}{8}$ and $\frac{1}{4}$ is 8, because 8 is a multiple of both 8 and ____.

4

120. A common denominator for the fractions $\frac{1}{3}$ and $\frac{1}{6}$ is 12, since 12 is a multiple of 3 and also a multiple of 6. However, 6 is also a common denominator of $\frac{1}{3}$ and $\frac{1}{6}$, because 6 is a multiple of both ____ and ____.

3; 6   (*Or* 6; 3.)

Remark. It is generally easiest to work with the smallest positive integer that is a multiple of the denominators of a set of fractions. The smallest such integer is called the **least common denominator** of the fractions.

121. Since 10 is the smallest positive integer that is a multiple of 2 and 5, 10 is the least ____ ____ of $\frac{1}{2}$ and $\frac{3}{5}$.

---

common denominator

122. Because 12 is the smallest positive integer that is a multiple of 3 and 12, 12 is the ____ ____ ____ of the fractions $\frac{1}{3}$ and $\frac{5}{12}$.

---

least common denominator

123. The least common denominator of the fractions $\frac{2}{7}$ and $\frac{2}{3}$ is ____ .

21

176 UNIT IV

Remark. Now, we shall represent a set of rational numbers by using fractions all of which have a common denominator. When we do this, we will be able to find a basic numeral or basic fraction for the sum of any rational numbers.

124. Recall that the Fundamental principle of fractions,

$$\frac{a}{b} = \frac{a \cdot c}{b \cdot c} \quad (b, c \neq 0),$$

asserts that the result of multiplying the numerator and denominator of a fraction by the same nonzero integer is a fraction that represents the same rational number as the original fraction. Thus, $\frac{3}{4} = \frac{3 \cdot 5}{4 \cdot 5} = \underline{\qquad}$.

$$\frac{3 \cdot 5}{4 \cdot 5} = \frac{15}{20}$$

125. $\frac{4}{5} = \frac{4 \cdot \underline{\quad}}{5 \cdot \underline{\quad}} = \frac{24}{30}$.

$$\frac{4 \cdot 6}{5 \cdot 6} = \frac{24}{30}$$

126. To obtain a factor that can be used to rewrite $\frac{3}{4}$ as a fraction having a denominator of 24, you think "What must 4 be multiplied by to give 24?" Since the answer to this question is 6, the desired factor is _____.

6

127. $\frac{3}{4} = \frac{3 \cdot 6}{4 \cdot 6} = \underline{\qquad}$.

$$\frac{18}{24}$$

128. The sum $\frac{1}{2} + \frac{1}{4}$ can be rewritten in a form in which both fractions have a common denominator. For example, $\frac{1}{2} + \frac{1}{4} = \frac{1 \cdot 2}{2 \cdot 2} + \frac{1}{4} = \frac{2}{4} + \frac{1}{4}$. This was accomplished by multiplying the numerator and denominator of $\frac{1}{2}$ by the factor ____ to change the fraction to $\frac{2}{4}$.

2

The Set of Rational Numbers 177

129. Since $\frac{2}{4} + \frac{1}{4} = \frac{3}{4}$, the sum $\frac{1}{2} + \frac{1}{4} = $ _____ .

$\frac{3}{4}$

130. Consider the sum $\frac{1}{3} + \frac{5}{8}$. The least common denominator of the fractions is _____ .

24

131. $\frac{1}{3} + \frac{5}{8}$ can be rewritten as $\frac{1 \cdot 8}{3 \cdot 8} + \frac{5 \cdot 3}{8 \cdot 3}$ by using the factors 8 and 3, respectively. The two fractions in the latter form now have the common denominator 24, and the sum can be expressed as $\frac{1 \cdot 8 + 5 \cdot 3}{3 \cdot 8}$ or by the basic fraction _____ .

$\frac{23}{24}$

132. $\frac{3}{7} + \frac{5}{9} = \frac{3 \cdot 9}{7 \cdot 9} + \frac{5 \cdot 7}{9 \cdot 7} = \frac{27 + 35}{63} = $ _____ .

$\frac{62}{63}$

133. $\frac{2}{3} + \frac{4}{5} = \frac{2 \cdot 5}{3 \cdot 5} + \frac{4 \cdot 3}{5 \cdot 3} = \frac{10 + 12}{15} = $ _____ .

$\frac{22}{15}$

134. $\frac{3}{4} + \frac{1}{8} = $ _____ .

$\frac{7}{8}$

Remark. As usual, we define

$$\frac{a}{b} + \frac{c}{d} + \frac{e}{f} = \left(\frac{a}{b} + \frac{c}{d}\right) + \frac{e}{f}$$

and use the Associative law as convenient.

UNIT IV

135. The least common denominator of the fractions in the sum $\frac{1}{2} + \frac{5}{6} + \frac{2}{15}$ is _____.

30

136. The sum $\frac{1}{2} + \frac{5}{6} + \frac{2}{15}$ can be rewritten as the sum _____ + _____ + _____, where each fraction has the denominator 30.

$\frac{15}{30} + \frac{25}{30} + \frac{4}{30}$ $\left(\frac{1 \cdot 15}{2 \cdot 15} + \frac{5 \cdot 5}{6 \cdot 5} + \frac{2 \cdot 2}{15 \cdot 2} \right)$

137. $\frac{1}{2} + \frac{5}{6} + \frac{2}{15}$, by definition, means $\left(\frac{1}{2} + \frac{5}{6} \right) + \frac{2}{15}$, or, when rewritten, $\left(\frac{15}{30} + \frac{25}{30} \right) + \frac{4}{30}$. This sum can then be represented by the basic fraction _____.

$\frac{22}{15}$ $\left(\frac{44}{30} \text{ will reduce to } \frac{22}{15}. \right)$

138. Write $\frac{5}{9} + \frac{3}{4} + \frac{5}{6}$ as a basic fraction.

$\frac{77}{36}$ $\left(\text{You first have } \frac{20}{36} + \frac{27}{36} + \frac{30}{36}. \right)$

Remark. Let us turn now to another way of expressing sums, namely by what are frequently called "mixed numbers" but are, in reality, **mixed numerals**.

139. Fractions are termed **proper** or **improper** according to whether the absolute value of the numerator is less or greater than the absolute value of the denominator, respectively. Thus, $\frac{1}{3}$ is a(n) _____ fraction, and $\frac{3}{1}$ is a(n) _____ fraction.

---

proper; improper

140. Since the numerator is greater than the denominator, $\frac{7}{4}$ is a(n) (proper/improper) fraction.

---

improper

The Set of Rational Numbers 179

141. Fractions with identical nonzero numerators and denominators are also improper fractions. Thus, $\frac{4}{4}$ is a(n) _____ fraction.

---

improper

142. Now, consider the sum of two positive numbers, one of which is a rational number that is also an integer, and one of which is a rational number that is not an integer, such as the sum $3 + \frac{1}{4}$. The symbol $3\frac{1}{4}$, read "three and one-fourth," is understood to represent the expression $3 + \frac{1}{4}$. Similarly, $2\frac{1}{3}$ represents

$2 + $ _____ .

$2 + \frac{1}{3}$

143. Symbols such as $4\frac{1}{2}$ and $2\frac{5}{6}$ are called mixed numerals. $5\frac{1}{4}$ is a _____ numeral that represents the _____ of 5 and $\frac{1}{4}$.

---

mixed; sum

144. The Closure law for addition in the system of rational numbers assures us that the sum of 5 and $\frac{1}{4}$ or $5\frac{1}{4}$ is a _____ number.

---

rational

145. Since 3, $\frac{3}{1}$, and $\frac{6}{2}$ all denote the same number, the sum $3 + \frac{1}{2}$ can be written as $\frac{3}{1} + \frac{1}{2} = \frac{2 \cdot 3}{2 \cdot 1} + \frac{1}{2} = \frac{6 + 1}{2} = $ _____ .

$\frac{7}{2}$

146. $4\frac{2}{3}$ can be written as $4 + \frac{2}{3}$. Since 4 can be represented by $\frac{4}{1} = \frac{3 \cdot 4}{3 \cdot 1} = \frac{12}{3}$,

▼

we have $4 + \frac{2}{3} = \frac{12}{3} + \frac{2}{3} = \frac{14}{3}$. The mixed numeral $4\frac{2}{3}$, and the improper fraction ____ , both represent the same number.

$\frac{14}{3}$

Remark. Any mixed numeral corresponds to an improper basic fraction. In any instance where an arithmetic expression such as a sum is written using mixed numerals, it can also be written using improper fractions. For example,

$$\begin{aligned}
2\frac{3}{4} + 4\frac{2}{3} &= \left(2 + \frac{3}{4}\right) + \left(4 + \frac{2}{3}\right) \\
&= \left(\frac{2 \cdot 4}{1 \cdot 4} + \frac{3}{4}\right) + \left(\frac{4 \cdot 3}{1 \cdot 3} + \frac{2}{3}\right) \\
&= \frac{11}{4} + \frac{14}{3} \\
&= \frac{11 \cdot 3}{4 \cdot 3} + \frac{14 \cdot 4}{3 \cdot 4} \\
&= \frac{33}{12} + \frac{56}{12} = \frac{89}{12}.
\end{aligned}$$

Let us go through a brief sequence of frames here to review some ideas concerning sums of rational numbers.

R147. If, $a, b, c \in J$ ($c \neq 0$), then $\frac{a}{c} + \frac{b}{c}$ can be written as the basic fraction ____ .

$\frac{a + b}{c}$

R148. The sum $\frac{2}{5} + \frac{1}{5}$ can be written as the basic fraction ____ .

$\frac{3}{5}$

R149. The sum $\frac{3}{5} + \frac{4}{7} = \frac{3 \cdot 7}{5 \cdot 7} + \frac{\cdot 4}{\cdot 7} = \frac{+}{35} = $ ____ .

$\frac{3 \cdot 7}{5 \cdot 7} + \frac{5 \cdot 4}{5 \cdot 7} = \frac{21 + 20}{35} = \frac{41}{35}$

The Set of Rational Numbers

R150. The sum $\dfrac{2}{3} + \dfrac{3}{4} = \dfrac{2 \cdot \underline{}}{3 \cdot \underline{}} + \dfrac{\underline{} \cdot 3}{\underline{} \cdot 4} = \dfrac{\underline{} + \underline{}}{3 \cdot 4} = \dfrac{\underline{}}{12}.$

$\dfrac{2 \cdot 4}{3 \cdot 4} + \dfrac{3 \cdot 3}{3 \cdot 4} = \dfrac{2 \cdot 4 + 3 \cdot 3}{3 \cdot 4} = \dfrac{17}{12}$

R151. The mixed numeral $5\dfrac{4}{5}$ represents the _____ of 5 and $\dfrac{4}{5}$.

---

sum

R152. $2\dfrac{1}{2} + 3\dfrac{1}{4} = \left(2 + \dfrac{1}{2}\right) + \left(3 + \dfrac{1}{4}\right) = \left(\dfrac{2 \cdot 2}{1 \cdot 2} + \dfrac{1}{2}\right) + \left(\dfrac{3 \cdot 4}{1 \cdot 4} + \dfrac{1}{4}\right)$

$= \dfrac{4+1}{2} + \dfrac{12+1}{4} = \dfrac{5 \cdot 2}{2 \cdot 2} + \dfrac{13}{4} = \dfrac{\underline{}}{4}.$

$\dfrac{10 + 13}{4} \quad \left(Or\ \dfrac{23}{4}.\right)$

Remark. See Exercise IVc, page 270, for additional practice in rewriting sums of rational numbers.

Now, we shall consider products of rational numbers. First, let us consider products of the form $\dfrac{1}{a} \cdot \dfrac{1}{b}$.

153. Recall that $3 \cdot \dfrac{1}{3} = 1$ and $5 \cdot \dfrac{1}{5} = 1$. Thus, $3 \cdot \dfrac{1}{3} \cdot 5 \cdot \dfrac{1}{5} = $ _____ .

$1 \quad (1 \cdot 1 = 1)$

154. By the Commutative law of multiplication and the Associative law of multiplication, $3 \cdot \dfrac{1}{3} \cdot 5 \cdot \dfrac{1}{5}$ can be written $3 \cdot 5 \cdot \dfrac{1}{3} \cdot \dfrac{1}{5}$. Since $3 \cdot \dfrac{1}{3} \cdot 5 \cdot \dfrac{1}{5} = 1$, then $3 \cdot 5 \cdot \dfrac{1}{3} \cdot \dfrac{1}{5}$ also equals _____ .

1

155. Now, $(3 \cdot 5) \cdot \left(\dfrac{1}{3} \cdot \dfrac{1}{5}\right) = 1$. For this assertion to obey our axiom, the factor

▼

$\left(\dfrac{1}{3} \cdot \dfrac{1}{5}\right)$ must be the multiplicative inverse of $(3 \cdot 5)$ or $\dfrac{1}{3 \cdot 5}$. Thus, we can write $\dfrac{1}{3} \cdot \dfrac{1}{5}$ as _____ .

$\dfrac{1}{3 \cdot 5}$ $\left(\text{Or } \dfrac{1}{15}\right)$

Remark. What we have done here is to show that $\dfrac{1}{3} \cdot \dfrac{1}{5} = \dfrac{1}{3 \cdot 5} = \dfrac{1}{15}$, where $\dfrac{1}{15}$ is a fraction representing the same rational number as the product $\dfrac{1}{3} \cdot \dfrac{1}{5}$. This example suggests the theorem:

Theorem: If $a, b \in J$ $(a, b \neq 0)$, then $\dfrac{1}{a} \cdot \dfrac{1}{b} = \dfrac{1}{a \cdot b}$.

Proof:

$a \cdot \dfrac{1}{a} = 1$ and $b \cdot \dfrac{1}{b} = 1$ (Multiplicative inverse axiom)

$\left(a \cdot \dfrac{1}{a}\right) \cdot \left(b \cdot \dfrac{1}{b}\right) = 1$ ($1 \cdot 1 = 1$; Substitution axiom)

$(a \cdot b) \cdot \left(\dfrac{1}{a} \cdot \dfrac{1}{b}\right) = 1$ (Associative and Commutative laws of multiplication)

$\dfrac{1}{a} \cdot \dfrac{1}{b} = \dfrac{1}{a \cdot b}$ (Multiplicative inverse axiom)

Logically, our next step is to see what can be done with products of the form $\dfrac{a}{b} \cdot \dfrac{c}{d}$, where a, b, c, and d are integers, except that b and d are not zero.

156. Consider the product $\dfrac{2}{3} \cdot \dfrac{4}{5}$.

$$\dfrac{2}{3} \cdot \dfrac{4}{5} = 2 \cdot \dfrac{1}{3} \cdot 4 \cdot \dfrac{1}{5}$$

$$= 2 \cdot 4 \cdot \dfrac{1}{3} \cdot \dfrac{1}{5}$$

$$= 2 \cdot 4 \cdot \dfrac{1}{3 \cdot 5} = \dfrac{2 \cdot 4}{3 \cdot 5}$$

$$= \underline{} .$$

$\dfrac{8}{15}$

The Set of Rational Numbers

Remark. The preceding example suggests the theorem:

Theorem: If $a, b, c, d \in J$ $(b, d \neq 0)$, then $\dfrac{a}{b} \cdot \dfrac{c}{d} = \dfrac{a \cdot c}{b \cdot d}$.

Proof:

$$\dfrac{a}{b} \cdot \dfrac{c}{d} = a \cdot \dfrac{1}{b} \cdot c \cdot \dfrac{1}{d} \qquad \left(\text{Theorem: } \dfrac{a}{b} = a \cdot \dfrac{1}{b}\right)$$

$$= a \cdot c \cdot \dfrac{1}{b} \cdot \dfrac{1}{d} \qquad \text{(Commutative and Associative laws of multiplication)}$$

$$= a \cdot c \cdot \dfrac{1}{b \cdot d} \qquad \left(\text{Theorem: } \dfrac{1}{b} \cdot \dfrac{1}{d} = \dfrac{1}{b \cdot d}\right)$$

$$= \dfrac{a \cdot c}{b \cdot d} \qquad \left(\text{Theorem: } a \cdot \dfrac{1}{b} = \dfrac{a}{b}\right)$$

157. Thus, $\dfrac{7}{9} \cdot \dfrac{1}{2}$ can be written as $\dfrac{7 \cdot 1}{9 \cdot 2}$ and then as the basic fraction _____.

$\dfrac{7}{18}$

158. $\dfrac{2}{3} \cdot \dfrac{11}{5} = $ _____.

$\dfrac{22}{15} \left(\text{Or } \dfrac{2 \cdot 11}{3 \cdot 5}\right)$

Remark. Here is what we now know about products of rational numbers:

If $a, b, c, d \in J$ $(a, b, d \neq 0)$, then

$$\dfrac{1}{a} \cdot \dfrac{1}{b} = \dfrac{1}{a \cdot b}, \text{ and } \dfrac{a}{b} \cdot \dfrac{c}{d} = \dfrac{a \cdot c}{b \cdot d}.$$

We obtained these relationships by applying the axioms we have adopted for the rational numbers. Because they were obtained through a series of abstractions, it might prove helpful to take a look at these products in the light of lattices that we used earlier in discussing the products of natural numbers.

159. Recall that we can associate a number, called the product of a and b, with the number of elements in the lattice or array containing a columns and b rows. The lattice

184   UNIT IV

has 5 columns, 3 rows, and _____ elements. The product 5 · 3 can be associated with the number of elements in the lattice and can be written as the basic numeral _____ .

15; 15

160. Using the same lattice, consider the number of elements enclosed in the dashed rectangle below. The rectangle is formed by taking 4 of the 5 columns $\left(\frac{4}{5}\right)$ and 2 of the 3 rows $\left(\frac{2}{3}\right)$.

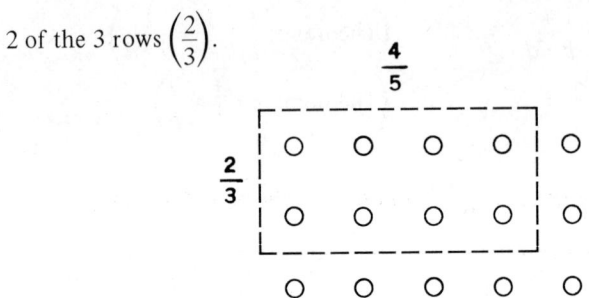

The number of elements in the dashed rectangle is _____ , and the number of elements in the entire lattice is _____ .

8; 15

161. The fact that there are 5 columns and 3 rows in the lattice

permits us to visualize the total number of elements in the lattice (15) as the product of 5 and 3. Similarly, the fact that there are 4 columns and 2 rows in the dashed rectangle permits us to visualize the total number of elements in the rectangle (8) as the product of _____ and _____ .

4; 2

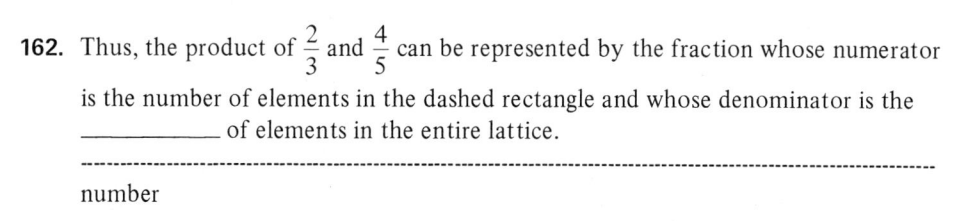

The Set of Rational Numbers   185

162. Thus, the product of $\frac{2}{3}$ and $\frac{4}{5}$ can be represented by the fraction whose numerator is the number of elements in the dashed rectangle and whose denominator is the _____ of elements in the entire lattice.

---

number

163. Consider the elements in the rectangle formed by taking only one of the 5 columns $\left(\frac{1}{5}\right)$ and only one of the two rows $\left(\frac{1}{2}\right)$ in the lattice shown:

The number of elements in the dashed rectangle is ____, from which we infer that the product $\frac{1}{5} \cdot \frac{1}{2}$ can be written as the basic fraction ____.

1; $\frac{1}{10}$

164. Construct a lattice to illustrate how the product $\frac{1}{7} \cdot \frac{1}{3}$ can be associated with the basic fraction $\frac{1}{21}$.

---

165. Construct a lattice to illustrate how the product $\frac{4}{7} \cdot \frac{1}{3}$ can be associated with the basic fraction $\frac{4}{21}$.

Remark. Does this approach to the product of two rational numbers seem more concrete? We are not proving that if $a, b, c, d \in J$ $(b, d \neq 0)$,

$$\frac{a}{b} \cdot \frac{c}{d} = \frac{a \cdot c}{b \cdot d}$$

through the use of a lattice, but it does illustrate the idea. Let us get on with some additional applications of this theorem.

166. If the fractions representing rational numbers in a product contain *common* factors in their numerators and denominators, the fraction resulting from applying

$$\frac{a}{b} \cdot \frac{c}{d} = \frac{a \cdot c}{b \cdot d}$$

will not be a basic fraction. For example, upon rewriting $\frac{3}{4} \cdot \frac{2}{5}$ as $\frac{3 \cdot 2}{4 \cdot 5}$, and then as $\frac{6}{20}$, we observe that the result, $\frac{6}{20}$, (is/is not) a basic fraction.

---

is not (6 and 20 contain the common factor 2.)

167. In general, we can obtain a basic fraction by

1. applying the relationship $\frac{a}{b} \cdot \frac{c}{d} = \frac{a \cdot c}{b \cdot d}$,
2. rewriting the products $a \cdot c$ and $b \cdot d$ in a completely factored form, and then
3. invoking the fundamental principle of fractions.

For example,

$$\frac{3}{8} \cdot \frac{10}{9} = \frac{3 \cdot 10}{8 \cdot 9}$$

$$= \frac{3 \cdot (2 \cdot 5)}{(2 \cdot 2 \cdot 2) \cdot (3 \cdot 3)}$$

$$= \frac{5 \cdot (2 \cdot 3)}{12 \cdot (2 \cdot 3)}$$

$$= \underline{}.$$

$\frac{5}{12}$

The Set of Rational Numbers 187

168. $\dfrac{3}{8} \cdot \dfrac{4}{27} = \dfrac{3 \cdot 4}{8 \cdot 27} = \dfrac{3 \cdot 2 \cdot 2}{2 \cdot 2 \cdot 2 \cdot 3 \cdot 3 \cdot 3}$

$= \dfrac{1 \cdot (3 \cdot 2 \cdot 2)}{18 \cdot (3 \cdot 2 \cdot 2)}$

$= \underline{}.$

$\dfrac{1}{18}$

169. Recall that we can use slant bars to indicate an application of the Fundamental principle of fractions. Thus, the example in the previous frame can be written

$$\dfrac{3}{8} \cdot \dfrac{4}{27} = \dfrac{\overset{1}{\cancel{3}} \cdot \overset{1}{\cancel{4}}}{\underset{2}{\cancel{8}} \cdot \underset{9}{\cancel{27}}} = \underline{}.$$

$\dfrac{1}{18}$ (*The numerators and denominators were first divided by 3, and then by 4.*)

170. Because the factors in the numerator and those in the denominator of $\dfrac{a \cdot c}{b \cdot d}$ are the same as those in the numerators and denominators, respectively, of $\dfrac{a}{b}$ and $\dfrac{c}{d}$, common factors in the numerators and denominators of $\dfrac{a}{b} \cdot \dfrac{c}{d}$ can be divided out prior to writing $\dfrac{a \cdot c}{b \cdot d}$. For example, we can write

$$\dfrac{\overset{1}{\cancel{9}}}{\underset{2}{\cancel{10}}} \cdot \dfrac{\overset{1}{\cancel{5}}}{\underset{2}{\cancel{18}}} = \dfrac{1}{4}$$

without first writing $\dfrac{9 \cdot 5}{10 \cdot 18}$. Similarly, the product of $\dfrac{5}{9}$ and $\dfrac{3}{10}$ can be written as a basic fraction as follows:

$$\dfrac{\overset{1}{\cancel{5}}}{\underset{3}{\cancel{9}}} \cdot \dfrac{\overset{1}{\cancel{3}}}{\underset{2}{\cancel{10}}} = \underline{}.$$

$\dfrac{1}{6}$

171. $\dfrac{3}{4} \cdot \dfrac{2}{21} = \dfrac{\cancel{3}^1}{\cancel{4}_2} \cdot \dfrac{\cancel{2}^1}{\cancel{21}_7} = $ _____ .

$\dfrac{1}{14}$

Remark. Theorems that are applicable to writing products of fractions as basic fractions *are not applicable* to products that contain mixed numerals. Recall, however, that any mixed numeral can be rewritten in the form of an improper fraction and therefore such products offer no particular difficulty.

172. $2\dfrac{1}{5} = 2 + \dfrac{1}{5} = \dfrac{10}{5} + \dfrac{1}{5} = $ _____ .

$\dfrac{11}{5}$

173. The product $3 \cdot 2\dfrac{1}{5}$ can be written as a basic fraction by first replacing the mixed numeral $2\dfrac{1}{5}$ with the improper fraction $\dfrac{11}{5}$. Hence, $3 \cdot 2\dfrac{1}{5} = \dfrac{3}{1} \cdot \dfrac{11}{5} = $ _____ .

$\dfrac{33}{5}$

174. $2 \cdot 4\dfrac{1}{3} = \dfrac{2}{1} \cdot \dfrac{13}{3} = $ _____ . Also, $5 \cdot 1\dfrac{2}{3} = $ _____ .

$\dfrac{26}{3}$; $\dfrac{25}{3}$ $\left(\text{You first have } 5 \cdot \dfrac{5}{3}. \right)$

Remark. This brings us to the end of our discussion of products of rational numbers. Let's review the ideas we have covered.

R175. If a and b are nonzero integers, $\dfrac{1}{a} \cdot \dfrac{1}{b} = $ _____ . Thus, $\dfrac{1}{9} \cdot \dfrac{1}{3} = $ _____ .

$\dfrac{1}{a} \cdot \dfrac{1}{b} = \dfrac{1}{a \cdot b}$; $\dfrac{1}{27}$

The Set of Rational Numbers 189

R176. If $a,b,c,d \in J$ ($b, d \neq 0$), then $\dfrac{a}{b} \cdot \dfrac{c}{d} = $ _____ . Thus, $\dfrac{5}{6} \cdot \dfrac{5}{7} = $ _____ .

$\dfrac{a}{b} \cdot \dfrac{c}{d} = \dfrac{a \cdot c}{b \cdot d}; \dfrac{25}{42}$

R177. Construct a lattice to illustrate how the product $\dfrac{2}{7} \cdot \dfrac{2}{3}$ can be associated with the basic fraction $\dfrac{4}{21}$.

R178. $3 \cdot 5\dfrac{2}{9}$ can be written as the basic fraction _____ .

$\dfrac{47}{3} \quad \left(\dfrac{3}{1} \cdot \dfrac{47}{9} = \dfrac{3 \cdot 47}{3 \cdot 3} = \dfrac{47}{3} \right)$

R179. $\dfrac{2}{7} \cdot 3\dfrac{1}{5}$ can be written as the basic fraction _____ .

$\dfrac{32}{35}$

Remark. See Exercise IVd, page 272, for additional practice on rewriting products of rational numbers.

We shall now take a brief look at the operations of subtraction and division in the system of rational numbers.

We can define the difference of two rational numbers in a manner similar to our definition of the difference of two integers, namely:

For all $\dfrac{a}{b}, \dfrac{c}{d} \in Q$ (*rational numbers*),

$$\dfrac{a}{b} - \dfrac{c}{d} = \dfrac{a}{b} + \left(-\dfrac{c}{d}\right).$$

▼

190   UNIT IV

Thus, an expression such as $\dfrac{8}{11} - \dfrac{3}{11}$ means $\dfrac{8}{11} + \left(-\dfrac{3}{11}\right)$.

In this example,
$$\dfrac{8}{11} - \dfrac{3}{11} = \dfrac{8}{11} + \left(-\dfrac{3}{11}\right) = \dfrac{8-3}{11} = \dfrac{5}{11}.$$

Observe that this definition is also consistent with the definition for the difference of two whole numbers, because $\dfrac{5}{11}$ is a number that added to $\dfrac{3}{11}$ equals $\dfrac{8}{11}$.

180. The difference $\dfrac{23}{3} - \dfrac{15}{3} = \dfrac{23}{3} + \left(-\dfrac{15}{3}\right) = \dfrac{23-15}{3} = \underline{}$.

$\dfrac{8}{3}$

181. $\dfrac{7}{13} - \dfrac{4}{13}$ can be written as $\dfrac{\underline{}}{13} = \underline{}$.

$\dfrac{7-4}{13} = \dfrac{3}{13}$

182. $5 - \dfrac{2}{3} = \dfrac{5}{1} - \dfrac{2}{3} = \dfrac{5 \cdot 3}{1 \cdot 3} - \dfrac{2}{3} = \dfrac{15}{3} - \dfrac{2}{3} = \dfrac{\underline{}}{3} = \underline{}$.

$\dfrac{15-2}{3} = \dfrac{13}{3}$

183. $4 - \dfrac{1}{5} = \dfrac{4}{1} - \dfrac{1}{5} = \dfrac{4 \cdot 5}{1 \cdot 5} - \dfrac{1}{5} = \dfrac{\underline{}}{5} = \underline{}$.

$\dfrac{20-1}{5} = \dfrac{19}{5}$

184. Write the difference $6 - \dfrac{3}{4}$ as a basic fraction.

$\dfrac{21}{4}$

The Set of Rational Numbers

185. The difference $\frac{3}{5} - \frac{2}{7}$ can first be written $\frac{3 \cdot 7}{5 \cdot 7} - \frac{2 \cdot 5}{7 \cdot 5}$, and then as the basic fraction _____ .

$\frac{11}{35}$

186. Write the difference $\frac{5}{2} - \frac{2}{11}$ as a basic fraction.

$\frac{51}{22}$

187. $7\frac{1}{5} - 6\frac{2}{3} = \frac{36}{5} - \frac{20}{3}$

$= \frac{36 \cdot 3}{5 \cdot 3} - \frac{20 \cdot 5}{3 \cdot 5}$

$= \frac{108}{15} - \frac{100}{15} = $ _____ .

$\frac{8}{15}$

Remark. You may have noticed that in every case in the last sequence of frames the numerator of the first fraction was always *larger* than the numerator of the second fraction when both fractions were rewritten with a common denominator. Now, if you recall our work with the number line, this means that the result of each of these subtractions would lie to the *right* of the origin, that is, on the *positive* side. For example, finding the difference $\frac{5}{8} - \frac{2}{8}$ on the number line would appear thus:

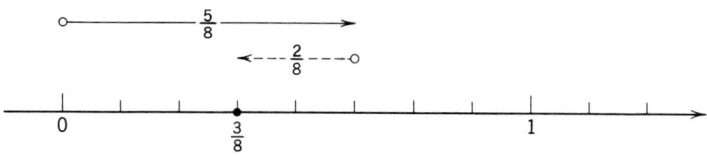

Since the set of rational numbers does include elements corresponding to points on the negative side of the number line, we can also find a difference such as $\frac{2}{8} - \frac{5}{8}$. This can be shown on the number line as follows:

▼

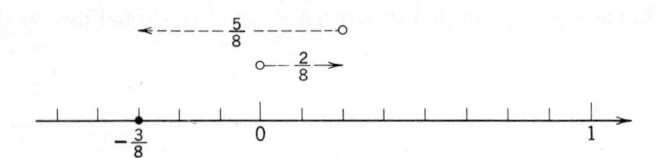

Our emphasis in this program however has been on positive rational numbers; we do not wish to dwell at length on negative differences.

As a last consideration in the system of rational numbers, we turn to the operation of division. We begin by reviewing briefly what is meant by the multiplicative inverse (or reciprocal) of a rational number.

188. Recall that, in the system of rational numbers, the multiplicative inverse of an integer a ($a \neq 0$) is the rational number $\frac{1}{a}$. Thus, the multiplicative inverse of 7 is $\frac{1}{7}$, and of 9 is _____ .

$\frac{1}{9}$

189. More generally, the multiplicative inverse of the rational number $\frac{a}{b}$ ($a, b \neq 0$) is the rational number $\frac{1}{\frac{a}{b}}$. The multiplicative inverse of $\frac{2}{3}$ is $\frac{1}{\frac{2}{3}}$, and of $\frac{3}{5}$ is _____ .

$\frac{1}{\frac{3}{5}}$

190. The product of a rational number and its multiplicative inverse is the identity element for multiplication, 1. Thus, $\frac{7}{8} \cdot \frac{1}{\frac{7}{8}} =$ _____ .

1

Remark. Now we wish to show that the multiplicative inverse of a rational number $\frac{a}{b}$ $\left(\frac{a}{b} \neq 0 \right)$ that we have been writing as $\frac{1}{\frac{a}{b}}$ can also be written as $\frac{b}{a}$. Let us consider a particular example.

The Set of Rational Numbers

191. Consider $\dfrac{1}{\frac{7}{8}}$, the reciprocal of $\dfrac{7}{8}$. By the fundamental principle of fractions, we may multiply both the numerator 1 and the denominator $\dfrac{7}{8}$ by $\dfrac{8}{7}$, and write

$$\frac{1}{\frac{7}{8}} = \frac{1 \cdot \left(\frac{8}{7}\right)}{\frac{7}{8} \cdot \left(\frac{8}{7}\right)}$$

Since $1 \cdot \dfrac{8}{7} = \dfrac{8}{7}$ and $\dfrac{7}{8} \cdot \dfrac{8}{7} = 1$, we have that $\dfrac{1}{\frac{7}{8}} = \dfrac{\frac{8}{7}}{1} = \underline{\phantom{\frac{8}{7}}}$.

$\dfrac{8}{7}$

Remark. The preceding example suggests the following theorem.

Theorem: If $\dfrac{a}{b} \in Q \left(\dfrac{a}{b} \neq 0\right)$, then $\dfrac{1}{\frac{a}{b}} = \dfrac{b}{a}$.

Proof:

$\dfrac{1}{\frac{a}{b}} = \dfrac{1 \cdot \frac{b}{a}}{\frac{a}{b} \cdot \frac{b}{a}}$ (Fundamental principle of fractions)

$\dfrac{1}{\frac{a}{b}} = \dfrac{\frac{b}{a}}{\frac{a}{b} \cdot \frac{b}{a}}$ (Multiplicative identity law)

$\dfrac{1}{\frac{a}{b}} = \dfrac{\frac{b}{a}}{1}$ $\left(\text{Theorem: } \dfrac{a}{b} \cdot \dfrac{c}{d} = \dfrac{a \cdot c}{b \cdot d}, \text{ and definition of quotient}\right)$

$\dfrac{1}{\frac{a}{b}} = \dfrac{b}{a}$ (Definition of quotient)

192. The reciprocal of $\frac{2}{3}$ is $\frac{1}{\frac{2}{3}}$ or $\frac{3}{2}$. The reciprocal of $\frac{3}{5}$ is $\frac{1}{\frac{3}{5}}$ or _____.

$\frac{5}{3}$

193. The reciprocals of 3, $\frac{1}{5}$, and $\frac{2}{7}$ can be written as the basic fractions (or numerals) ____, ____, and ____, respectively.

$\frac{1}{3}$; 5; $\frac{7}{2}$

Remark. Now, having established that the reciprocal or multiplicative inverse of a given rational number $\frac{a}{b}$ ($a \neq 0$) is $\frac{b}{a}$, let us see how we can use this idea to help us to rewrite expressions involving quotients of rational numbers. We shall see why the quotient $\frac{\frac{a}{b}}{\frac{c}{d}}$ or $\frac{a}{b} \div \frac{c}{d}$ can be written as $\frac{a}{b} \cdot \frac{d}{c}$, the product of the numerator and the reciprocal of the denominator. First, let us consider a numerical example.

194. Consider the quotient $\frac{\frac{2}{3}}{\frac{7}{5}}$ or $\frac{2}{3} \div \frac{7}{5}$. By the fundamental principle of fractions, we may write $\frac{\frac{2}{3}}{\frac{7}{5}} = \frac{\frac{2}{3} \cdot \frac{5}{7}}{\frac{7}{5} \cdot \frac{5}{7}}$.

Since $\frac{7}{5} \cdot \frac{5}{7} = 1$, we have $\frac{\frac{2}{3}}{\frac{7}{5}} = \frac{\frac{2}{3} \cdot \frac{5}{7}}{1}$

and we see that $\frac{\frac{2}{3}}{\frac{7}{5}}$ or $\frac{2}{3} \div \frac{7}{5}$ can be written as the product $\frac{2}{3} \cdot$ _____.

$\frac{2}{3} \cdot \frac{5}{7}$

The Set of Rational Numbers 195

195. In the preceding frame we observed that $\frac{2}{3} \div \frac{7}{5} = \frac{2}{3} \cdot \frac{5}{7}$. The right-hand member of this equation is the product of $\frac{2}{3}$ and $\frac{5}{7}$, the reciprocal or multiplicative inverse of ____ .

$\frac{7}{5}$

Remark. The preceding frames suggest the following theorem.

Theorem: If $\frac{a}{b}, \frac{c}{d} \in Q \left(\frac{c}{d} \neq 0\right)$ then $\frac{a}{b} \div \frac{c}{d} = \frac{a}{b} \cdot \frac{d}{c}$

Proof:

$$\frac{\frac{a}{b}}{\frac{c}{d}} = \frac{\frac{a}{b} \cdot \frac{d}{c}}{\frac{c}{d} \cdot \frac{d}{c}} \qquad \text{(Fundamental principle of fractions)}$$

$$\frac{\frac{a}{b}}{\frac{c}{d}} = \frac{\frac{a}{b} \cdot \frac{d}{c}}{1} \qquad \left(\text{Theorem: } \frac{a}{b} \cdot \frac{c}{d} = \frac{a \cdot c}{b \cdot d}, \text{ and definition of quotient}\right)$$

$$\frac{\frac{a}{b}}{\frac{c}{d}} = \frac{a}{b} \div \frac{c}{d} = \frac{a}{b} \cdot \frac{d}{c} \qquad \text{(Definition of quotient)}$$

196. The multiplicative inverse of $\frac{5}{2}$ is ____ . The quotient $\frac{3}{4} \div \frac{5}{2}$ can be written $\frac{3}{4} \cdot \frac{2}{5}$, the ____ of $\frac{3}{4}$ and $\frac{2}{5}$.

---

product; $\frac{2}{5}$

197. The quotient $\frac{2}{5} \div \frac{4}{7}$ can be written as the product $\frac{2}{5} \cdot$ ____ .

$\frac{2}{5} \cdot \frac{7}{4}$

196 UNIT IV

198. The quotient of $\frac{a}{b}$ divided by $\frac{c}{d}$, is equal to the product of $\frac{a}{b}$ and the _____ _____ of $\frac{c}{d}$.

multiplicative inverse *(Or reciprocal.)*

199. The quotient $\frac{3}{4} \div \frac{2}{5} =$ _____ · _____ = _____ .

$\frac{3}{4} \cdot \frac{5}{2}; \frac{15}{8}$

Remark. Here we have the traditional statement "To divide one fraction by another, 'invert' the divisor and multiply." In truth, we are dividing rational numbers and not fractions, and, what's more, the quotient can be denoted simply by $\frac{a}{b} \div \frac{c}{d}$ or $\frac{\frac{a}{b}}{\frac{c}{d}}$. What we are doing in "inverting and multiplying" is transforming one representation of the quotient into another form, that of a basic fraction.

200. Because the set of rational numbers is *closed* for multiplication, and because the quotient of two rational numbers $\left(\frac{a}{b} \div \frac{c}{d}, \frac{c}{d} \neq 0\right)$ is *always* expressible as the product of rational numbers $\left(\frac{a}{b} \cdot \frac{d}{c}\right)$, the set of rational numbers is _____ for division, except for division by 0.

closed

201. If $\frac{a}{b}$ and $\frac{c}{d}$ are rational numbers $\left(\frac{c}{d} \neq 0\right)$, the quotient $\frac{a}{b} \div \frac{c}{d}$ *always* denotes a _____ number.

rational

202. The quotient $\frac{3}{4} \div \frac{2}{3}$ equals the product of $\frac{3}{4}$ and the _____ of $\frac{2}{3}$.

reciprocal *(Or multiplicative inverse.)*

The Set of Rational Numbers 197

203. Write $\frac{3}{7} \div \frac{2}{7}$ as an equal expression in the form of a product.

$\frac{3}{7} \cdot \frac{7}{2}$

204. Write the product $\frac{3}{7} \cdot \frac{7}{2}$ as a basic fraction.

$\frac{3}{2}$

205. The quotient $5 \div \frac{2}{3} = \frac{5}{1} \div \frac{2}{3} = \frac{5}{1} \cdot \frac{3}{2} = $ _____ .

$\frac{15}{2}$

206. The quotient $\frac{1}{5} \div 4 = \frac{1}{5} \div \frac{4}{1} = \frac{1}{5} \cdot \frac{1}{4} = $ _____ .

$\frac{1}{20}$

207. The quotient $6 \div \frac{3}{4} = $ _____ ; $7 \div \frac{2}{3} = $ _____ .

$8; \frac{21}{2}$

208. The quotient $\frac{4}{7} \div 7 = $ _____ ; $\frac{3}{5} \div 6 = $ _____ .

$\frac{4}{49}; \frac{1}{10}$

Remark. The next sequence of frames will give you an opportunity to review the work on the difference and quotient of two rational numbers. In the following frames assume that a, b, c, and d are integers and that no denominator equals 0.

R209. If $\frac{a}{b}$ and $\frac{c}{d}$ are rational numbers, then $\frac{a}{c} - \frac{b}{c} = $ _____ .

$\frac{a-b}{c}$

UNIT IV

R210. The difference $\dfrac{4}{7} - \dfrac{1}{7} = \dfrac{\rule{1cm}{0.4pt}}{7} = $ _____ .

$\dfrac{4-1}{7} = \dfrac{3}{7}$

R211. The difference $5 - \dfrac{2}{9} = \dfrac{5 \cdot 9}{9} - \dfrac{2}{9} = $ _____ .

$\dfrac{43}{9}$

R212. The difference $\dfrac{16}{3} - \dfrac{22}{5} = \dfrac{16 \cdot \rule{0.5cm}{0.4pt}}{3 \cdot \rule{0.5cm}{0.4pt}} - \dfrac{\rule{0.5cm}{0.4pt} \cdot 22}{\rule{0.5cm}{0.4pt} \cdot 5} = \dfrac{\rule{0.5cm}{0.4pt}}{15} = $ _____ .

$\dfrac{16 \cdot 5}{3 \cdot 5} - \dfrac{3 \cdot 22}{3 \cdot 5} = \dfrac{80 - 66}{15} = \dfrac{14}{15}$

R213. $\dfrac{1}{\frac{a}{b}}$ is called the _____ _____ or the _____ of $\dfrac{a}{b}$.

multiplicative inverse; reciprocal

R214. $\dfrac{1}{\frac{a}{b}} = 1 \cdot \dfrac{b}{a} = $ _____ .

$\dfrac{b}{a}$

R215. For $a, b \neq 0$, the reciprocal of $\dfrac{a}{b}$ is ____; the reciprocal of $\dfrac{5}{6}$ is ____.

$\dfrac{b}{a}$; $\dfrac{6}{5}$ $\left(Or\ \dfrac{1}{\frac{a}{b}};\ \dfrac{1}{\frac{5}{6}} \right)$

R216. The quotient $\dfrac{a}{b} \div \dfrac{c}{d}$ can be expressed as the product _____ .

$\dfrac{a}{b} \cdot \dfrac{d}{c}$ $\left(Or\ \dfrac{a \cdot d}{b \cdot c} \right)$

The Set of Rational Numbers

R217. The quotient $\frac{3}{4} \div \frac{2}{7}$ can be written as the product _____ · _____, and then as the basic fraction _____.

$\frac{3}{4} \cdot \frac{7}{2}; \frac{21}{8}$

R218. The quotient $\frac{3}{5} \div \frac{4}{7}$ can be written as the basic fraction _____.

$\frac{21}{20}$

R219. Since the quotient of two rational numbers $\left(\frac{a}{b} \div \frac{c}{d}\right)$ is a rational number, the set of rational numbers is _____ for division, except $\frac{c}{d} \neq 0$.

---

closed

Remark. See Exercise IVe, page 274, for additional practice on differences and quotients of rational numbers.

The following panel on page 201 summarizes the axioms and theorems for the rational numbers that have been introduced in this part of the program.

Panel III

$a, b, c, d \in J$

AXIOM

For all $\dfrac{a}{b} \neq 0$, there is exactly one number

$\dfrac{1}{\frac{a}{b}}$ with the property that $\dfrac{a}{b} \cdot \dfrac{1}{\frac{a}{b}} = 1.$ (Multiplicative inverse law)

PROPERTIES

$\dfrac{a}{b} = a \cdot \dfrac{1}{b} \quad (b \neq 0)$

$\dfrac{a}{b} = \dfrac{c}{d}$ if $a \cdot d = b \cdot c \quad (b, d \neq 0)$

$a \cdot d = b \cdot c$ if $\dfrac{a}{b} = \dfrac{c}{d} \quad (b, d \neq 0)$

$\dfrac{a \cdot c}{b \cdot c} = \dfrac{a}{b} \quad (b, c \neq 0)$

$\dfrac{a}{c} + \dfrac{b}{c} = \dfrac{a+b}{c} \quad (c \neq 0)$

$\dfrac{a}{c} - \dfrac{b}{c} = \dfrac{a-b}{c} \quad (c \neq 0)$

$\dfrac{1}{a} \cdot \dfrac{1}{b} = \dfrac{1}{a \cdot b} \quad (a, b \neq 0)$

$\dfrac{a}{b} \cdot \dfrac{c}{d} = \dfrac{a \cdot c}{b \cdot d} \quad (b, d \neq 0)$

$\dfrac{1}{\frac{a}{b}} = \dfrac{b}{a} \quad (a, b \neq 0)$

$\dfrac{a}{b} \div \dfrac{c}{d} = \dfrac{a}{b} \cdot \dfrac{d}{c} \quad (b, c, d \neq 0)$

The Set of Rational Numbers

Remark. Let us now summarize what we have learned about rational numbers up to this point in this part of the program. You may want to refer to Panel III as you complete the following frames.

R220. Rational numbers are quotients of _____.

---
integers

R221. A rational number can be represented by a *symbol* of the form $\frac{a}{b}$, called a _____.

---
fraction

R222. $\{1, 2, 3, 4, \ldots\}$ (is/is not) a subset of $\{$rational numbers$\}$.

---
is

R223. $\{\ldots -3, -2, -1, 0, 1, 2, 3 \ldots\}$ (is/is not) a subset of $\{$rational numbers$\}$.

---
is

R224. $\{$rational numbers$\}$ is a(n) (finite/infinite) set.

---
infinite

R225. If b is any nonzero integer, $\frac{1}{b}$ is the rational number such that $b \cdot \frac{1}{b} =$ _____.

$b \cdot \frac{1}{b} = 1$

R226. The product of two numbers, each of which is the multiplicative inverse, or _____, of the other is 1.

---
reciprocal

R227. The multiplicative inverse, or reciprocal, of 5 is ____.

$\frac{1}{5}$

R228. The axioms adopted for {integers} (are/are not) valid for {rational numbers}.

---
are

R229. To characterize the behavior of the new elements of the enlarged set of rational numbers, it is assumed that:

For every nonzero rational number $\dfrac{a}{b}$, there is exactly one number $\dfrac{1}{\frac{a}{b}}$, such that $\dfrac{a}{b} \cdot \dfrac{1}{\frac{a}{b}} = 1$.

This statement is called the _____ inverse axiom.

Multiplicative

R230. The axioms for the rational number system differ from the axioms for the system of integers only in that the rational number system includes the _____ _____ axiom.

Multiplicative inverse

R231. The rational number $\dfrac{a}{b}$ (always/sometimes/never) equals $a \cdot \dfrac{1}{b}$.

---
always

R232. If $a,b,c,d \in J$ $(b,d \neq 0)$, then $\dfrac{a}{b} = \dfrac{c}{d}$ if _____ $\cdot$ _____ = _____ $\cdot$ _____ .

$a \cdot d = c \cdot b$

R233. $\dfrac{3}{7} = \dfrac{6}{14}$ because _____ $\cdot$ _____ = _____ $\cdot$ _____ .

$3 \cdot 14 = 6 \cdot 7$

R234. One fraction can be written as another fraction representing the same rational number by invoking the _____ _____ of fractions, which states:

If $a, b, c \in J$ $(b,c \neq 0)$, then $\dfrac{a}{b} = \dfrac{a \cdot c}{b \cdot c}$.

Fundamental principle

The Set of Rational Numbers   205

R235. The fraction $\frac{12}{13}$ can be changed to $\frac{36}{39}$, which represents the same rational number, by _____ both numerator and denominator by ____ .

---

multiplying; 3

R236. If the numerator and denominator of $\frac{24}{18}$ are _____ by 6, the fraction $\frac{24}{18}$ can be written as ____ .

---

divided; $\frac{4}{3}$

R237. The fraction $\frac{a \cdot c}{b \cdot c}$ is said to be in _____ terms than $\frac{a}{b}$, and $\frac{a}{b}$ is said to be in _____ terms than $\frac{a \cdot c}{b \cdot c}$.

---

higher; lower

R238. If $a, b, c \in J$ $(c \neq 0)$, then $\frac{a}{c} + \frac{b}{c} = $ _____ .

$\frac{a}{c} + \frac{b}{c} = \frac{a+b}{c}$

R239. The sum of $\frac{3}{7}$ and $\frac{2}{7}$ can be written as the basic fraction ____ .

$\frac{5}{7}$

R240. If $a, b, c, d \in J$ $(b, d \neq 0)$, then

$$\frac{a}{b} + \frac{c}{d} = \frac{(d) \cdot a}{(d) \cdot b} + \frac{c \cdot (b)}{d \cdot (b)} = \frac{\underline{\quad} \cdot \underline{\quad} + \underline{\quad} \cdot \underline{\quad}}{\underline{\quad} \cdot \underline{\quad}}.$$

$\frac{a \cdot d + b \cdot c}{d \cdot b} \quad \left(\text{Or } \frac{d \cdot a + c \cdot b}{d \cdot b} \right)$

R241. $\frac{1}{2} + \frac{5}{6}$ can be written as the basic fraction ____ .

$\frac{4}{3}$

206   UNIT IV

R242. A mixed numeral, such as $7\frac{1}{9}$, is understood to denote the _____ of the rational numbers 7 and $\frac{1}{9}$. Hence, $7\frac{1}{9} = 7 + \frac{1}{9} = \frac{7 \cdot 9}{1 \cdot 9} + \frac{1}{9} = \frac{}{9}$.

---

sum; $\frac{64}{9}$

R243. $7\frac{1}{9} + \frac{4}{9} =$ _____ .

$\frac{68}{9}$

R244. If a and b are nonzero integers, $\frac{1}{a} \cdot \frac{1}{b} =$ _____ .

$\frac{1}{a} \cdot \frac{1}{b} = \frac{1}{a \cdot b}$

R245. $\frac{1}{5} \cdot \frac{1}{7}$ can be written as the basic fraction _____ .

$\frac{1}{35}$

R246. If $a, b, c, d \in J$ $(b, d \neq 0)$, then $\frac{a}{b} \cdot \frac{c}{d} =$ _____ .

$\frac{a}{b} \cdot \frac{c}{d} = \frac{a \cdot c}{b \cdot d}$

R247. $\frac{7}{3} \cdot \frac{5}{11}$ can be written as the basic fraction _____ .

$\frac{35}{33}$

R248. A product of two rational numbers can be represented through the use of a lattice. Recall that the lattice

▼

The Set of Rational Numbers 207

has 2 rows, 6 columns, and represents the product _____.

---
2 · 6 (Or 6 · 2 or 12.)

R249. The lattice represents the product $\frac{2}{3}$ · ___, which can be written as the basic fraction ___.

$\frac{4}{5}$, $\frac{8}{15}$

R250. Construct a lattice to illustrate that the product $\frac{3}{4} \cdot \frac{3}{5}$ can be written as the basic numeral $\frac{9}{20}$.

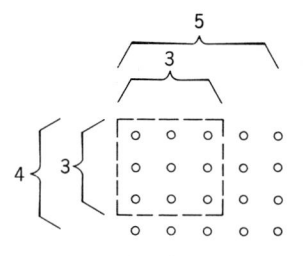

R251. The difference of two rational numbers is defined as follows:

For all $\frac{a}{b}, \frac{c}{d} \in Q$, $\frac{a}{b} - \frac{c}{d} = \frac{a}{b} + \Big(\quad\Big)$.

$\frac{a}{b} + \left(-\frac{c}{d}\right)$

R252. Express the difference $\frac{5}{7} - \frac{2}{7}$ as a basic fraction.

$\frac{3}{7}$

R253. The multiplicative inverse of any rational number $\frac{a}{b}$ $\left(\frac{a}{b} \neq 0\right)$ is equal to _____.

$\frac{b}{a}$ $\left(Or \ \frac{1}{\frac{a}{b}}\right)$

R254. The quotient, $\frac{a}{b} \div \frac{c}{d}$, equals the product of $\frac{a}{b}$ and the _____ _____ of $\frac{c}{d}$.

---

multiplicative inverse (*Or reciprocal.*)

R255. $\frac{3}{4} \div \frac{2}{3} = \frac{3}{4} \cdot$ _____ , which can be written as the basic fraction _____ .

$\frac{3}{2}$; $\frac{9}{8}$

R256. The quotient of two rational numbers $\frac{a}{b} \div \frac{c}{d}$, $\left(\frac{c}{d} \neq 0\right)$ is always expressible as a product of rational numbers $\frac{a}{b} \cdot \frac{d}{c}$. Because the set of rational numbers is closed for multiplication, it is also _____ for division, except division by _____ .

---

closed; zero

Remark. Up to this point we have considered several concepts concerning the set $Q = \{\text{rational numbers}\} = \{\frac{a}{b} | a, b \in J, \ b \neq 0\}$. The elements of the set of rational numbers were characterized by the axiom adopted for the system (see Panel III), and the structure of the system developed as we stated certain theorems and demonstrated their validity as logical consequences of our axioms.

Throughout this part of the program we have only used fraction symbols for rational numbers. You probably recall that other symbols such as decimal fractions (or simply decimals) are often used to represent rational numbers. We shall not consider this notation since our basic objective in this part of the program, which is to introduce the notion of the rational number system, is best accomplished with the fractional representations that we have used.

This concludes the study of the system of rational numbers. Try the self-evaluation test that follows. Score your paper from the answers given on page 279.

UNIT IV
Self-Evaluation Test

1. The number $\frac{1}{3}$ is called the _____ _____ of 3.

2. $\{\ldots -3, -2, -1, 0, 1, 2, 3, \ldots\}$ (is/is not) a subset of $\{\text{rational numbers}\}$.

3. The symbol 53 (does/does not) represent a rational number.

4. If $a, b \in J$ ($b \neq 0$), then $\frac{a}{b}$ (always/sometimes/never) equals $a \cdot \frac{1}{b}$.

5. Graph the rational number $\frac{5}{3}$ on a number line.

6. If $a, b, c, d \in J$ ($b, d \neq 0$) and if $\frac{a}{b} = \frac{c}{d}$, then $a \cdot d$ (always/sometimes/never) equals $b \cdot c$.

7. $\frac{5}{7}$ (does/does not) equal $\frac{13}{15}$.

8. The fraction $\frac{24}{30}$ can be reduced to the basic fraction _____.

9. Write $\frac{3}{5} + \frac{2}{7}$ as a single fraction.

10. $\frac{2}{7} + \frac{1}{3} + \frac{1}{21} =$ _____.

11. $\frac{1}{5} \cdot \frac{1}{3} =$ _____.

12. If $a, b, c, d \in J$ ($b, d \neq 0$), then $\frac{a}{b} \cdot \frac{c}{d}$ (always/sometimes/never) is equal to $\frac{a \cdot c}{b \cdot d}$.

13. $\frac{3}{8} \cdot \frac{4}{9} =$ _____.

14. Construct a lattice to illustrate the product $\frac{2}{5} \cdot \frac{3}{7}$.

15. $\dfrac{3}{5} - \dfrac{1}{3} = $ _____ .

16. If $\dfrac{a}{b} \in Q \left(\dfrac{a}{b} \neq 0\right)$, then $\dfrac{1}{\frac{a}{b}}$ (always/sometimes/never) is equal to $\dfrac{b}{a}$.

17. If $\dfrac{a}{b}, \dfrac{c}{d} \in Q \left(\dfrac{c}{d} \neq 0\right)$, then $\dfrac{a}{b} \div \dfrac{c}{d}$ is equal to the product _____ .

18. $\dfrac{7}{5} \div \dfrac{3}{4} = $ _____ .

19. $2\dfrac{1}{2} \div 3\dfrac{1}{4} = $ _____ .

20. If $\dfrac{a}{b}, \dfrac{c}{d} \in Q \left(\dfrac{c}{d} \neq 0\right)$, then $\dfrac{a}{b} \div \dfrac{c}{d}$ (always/sometimes/never) represents a rational number.

If you missed fewer than seven questions on this test, continue on to Unit V of this program. If you missed seven or more questions, you would probably profit by returning to the remark on page 147 and reading through the program to this point.

UNIT V
The Set of Real Numbers

OBJECTIVES

Upon completion of this part of the program, the reader will be able to:

1. Use the correct vocabulary and/or the correct symbolism to express square root, radical expression, irrational number, and the completeness property.

2. Express relationships between the sets of real numbers, rational numbers, irrational numbers, integers, whole numbers, and natural numbers.

3. Associate real numbers with points on a number line and the set of irrational numbers with a subset of the set of real numbers.

4. Associate the graphs of some infinite sets of real numbers with line segments on a number line.

5. Identify and illustrate the axioms and theorems for the real number system, and identify when each is being invoked.

Remark. In Unit IV of this program we noted that the set of rational numbers is infinite and their graphs are an infinite set of points on the number line. However, these numbers do not completely fill the number line. There are points on the number line that do not correspond to rational numbers. In this part of the program we shall consider numbers that can be associated with points on the number line and that are not rational. First, however, we will introduce another representation for a rational number which will also be used to represent the new numbers later in the program.

1. The **square root** of a number is one of two equal factors of the number. Because 3 is one of two equal factors of 9 (3 · 3 = 9), the number 3 is the square _____ of 9.

   ---
   root

2. Every positive number has two square roots. The number 5 is the positive _____ _____ of 25 because (5)(5) = 25 and −5 is the negative _____ _____ of 25 because (−5)(−5) = 25.

   ---
   square root; square root

3. The number 4 is the _____ square root of 16.

   ---
   positive

4. The positive square root of a positive number, say 16, can be represented by the symbol $\sqrt{16}$ (read "the square root of 16"). $\sqrt{16}$ represents 4 because 4 · 4 = 16. $\sqrt{9}$ represents _____ because 3 · 3 = 9. $\sqrt{25}$ = _____ because 5 · 5 = 25.

   3; 5

5. Since $\sqrt{16}$ equals 4 and $\sqrt{25}$ equals 5, $-\sqrt{16}$ represents −4 and $-\sqrt{25}$ represents _____.

   −5

6. $\sqrt{\frac{4}{9}} = \frac{2}{3}$ because $\frac{2}{3} \cdot \frac{2}{3} = \frac{4}{9}$. $\sqrt{\frac{4}{25}} =$ _____ because _____ · _____ = $\frac{4}{25}$.

   $\frac{2}{5}$; $\frac{2}{5} \cdot \frac{2}{5}$

214 UNIT V

7. $\sqrt{\frac{1}{4}} = \underline{}$; $-\sqrt{\frac{1}{4}} = \underline{}$.

$\frac{1}{2}$; $-\frac{1}{2}$

8. Expressions such as $\sqrt{144}$ and $\sqrt{100}$ are called **radical expressions** and the symbol "$\sqrt{}$" is called a **radical sign**. The number 10 can be written as the radical expression $\sqrt{100}$. The number 3 can be written as the radical expression $\underline{}$.

$\sqrt{9}$

9. 7 can be written as the radical expression $\underline{}$.

$\sqrt{49}$

10. -4 can be written as $-\sqrt{16}$ and -7 can be written as $\underline{}$.

$-\sqrt{49}$ ($\sqrt{-49}$ *is not the same thing.*)

11. $\frac{3}{4}$ can be written as $\sqrt{\frac{9}{16}}$ and $\frac{2}{3}$ can be written as $\underline{}$.

$\sqrt{\frac{4}{9}}$

12. In the radical expression $\sqrt{b}$, b is called the **radicand**. Thus, in the expression $\sqrt{25}$, the symbol "$\sqrt{}$" is called a radical sign and 25 is called the $\underline{}$.

---

radicand

13. Because $\sqrt{4} = 2$ and $2 \cdot 2 = 4$, it follows that $\sqrt{4}$ is a number such that $\sqrt{4} \cdot \sqrt{4} = 4$. Similarly, $\sqrt{25}$ is a number such that $\sqrt{25} \cdot \sqrt{25} = \underline{}$.

---

25 (5 · 5 = 25)

Remark. In general, we make the following definition:

$\sqrt{b}$, $b \geq 0$, is the non negative number such that $\sqrt{b} \cdot \sqrt{b} = b$.

The Set of Real Numbers

14. From the definition of square root, $\sqrt{36} \cdot \sqrt{36} =$ _____.

---
36 ($6 \cdot 6 = 36$)

15. $\sqrt{64} =$ _____; $\sqrt{81} =$ _____; $\sqrt{100} =$ _____.

8; 9; 10

Remark. In our work, we will limit our study to the square roots of nonnegative numbers. Therefore, radicands will always be nonnegative and wherever variables occur as radicands it will be assumed that the variables represent nonnegative numbers only. Thus, if we write $\sqrt{b}$, it is implied that b represents a positive number or zero.

We have now seen how radical notation can be used to represent rational numbers. In the following frames we will see how this notation can be used to represent a new kind of number.

16. Numbers that can be associated with points on a number line and which cannot be expressed in the form $\frac{a}{b}$, where $a,b \in J$, are called **irrational numbers**. For example, $\sqrt{2}$ is such a number; when it is multiplied by itself the product equals 2. Similarly $\sqrt{3}$ cannot be expressed in the form $\frac{a}{b}$, where $a,b \in J$, and hence, it is an _____ number.

---
irrational

17. In general, if the radicand in a radical expression is positive and is not the product of two equal rational numbers, then the radical expression represents an irrational number. For example, $\sqrt{16}$ is a rational number because $16 = 4 \cdot 4$. Because $\frac{16}{25} = \frac{4}{5} \cdot \frac{4}{5}$, $\sqrt{\frac{16}{25}}$ is a _____ number. Assuming that no rational number exists which when multiplied by itself equals 5, then $\sqrt{5}$ is an _____ number. Furthermore $\sqrt{5} \cdot \sqrt{5} =$ _____.

---
rational; irrational; 5

18. Assuming that no rational number exists, which when multiplied by itself equals 13, then $\sqrt{13}$ is an _____ number. Furthermore, $\sqrt{13} \cdot \sqrt{13} =$ _____.

---
irrational; 13

Remark. Some irrational numbers are not expressed in radical notation. For example, you probably recall that the quotient of the length of the circumference of a circle and the length of its diameter is also irrational and is designated by the Greek letter π (pi). Our concern at the moment is with irrational numbers that are expressed in radical form.

Although the irrational number $\sqrt{2}$ cannot be represented as the quotient of two integers, we can obtain a rational number approximation for this number to any degree of accuracy that we wish. There are several ways to do this. We will use the table on page 241 which gives rational number approximations of irrational numbers to three decimal places. We shall use the symbol $\approx$ to mean "is approximately equal to."

For example, from the table on page 241, we have that $\sqrt{2} \approx 1.414$ and $\sqrt{3} \approx 1.732$.

We have already seen how the rational numbers can be represented as points on a number line. Irrational numbers are also associated with points on the same number line used for rational numbers. In other words, the points on the number line are the graphs of the elements in both the set of rational numbers and the set of irrational numbers. The union of these sets is called the set of **real numbers**, R. We shall designate the set of irrational numbers by H. Thus, $Q \cup H = R$.

19. Rational numbers such as $-\sqrt{4}, \frac{4}{9}, \sqrt{9}$ can be associated with points on a line.

Their graphs appear as

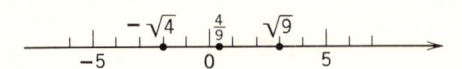

Graph $\{-\sqrt{16}, -\frac{4}{9}, \sqrt{25}\}$ on the line below.

---

20. Irrational numbers can also be associated with points on a line. The decimal approximations on page 241 can be used to *approximate* the position of their graphs on the line. Thus, $-\sqrt{2} \approx -1.414$ and $\sqrt{3} \approx 1.732$ can be graphed as

▼

The Set of Real Numbers 217

Approximate the position of the graphs of $-\sqrt{5}$ and $\sqrt{7}$ on the line below.

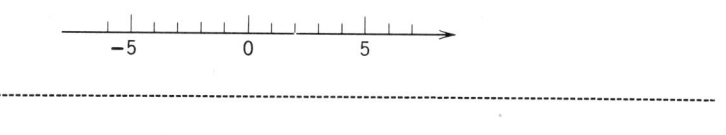

```
  |__|__|__|__|__|__|__|__|__|__|__|__|
         -5         0         5
```

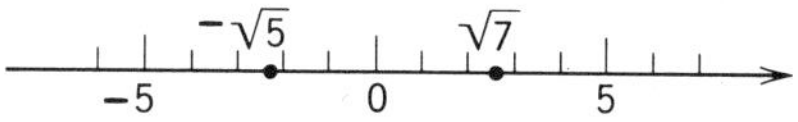

```
  |__|__|•__|__|__|__|__|•__|__|__|__|
        -√5         √7
         -5         0         5
```
$(-\sqrt{5} \approx -2.236;\ \sqrt{7} \approx 2.646)$

21. Graph $\{-\sqrt{10}, \sqrt{9},$ and $\sqrt{23}\}$.

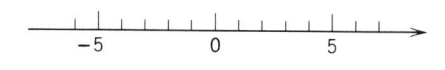

```
  |__|__|__|__|__|__|__|__|__|__|__|__|
         -5         0         5
```

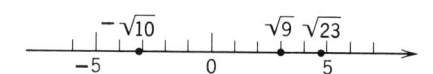

```
  |__|•__|__|__|__|__|__|•_•|__|__|__|
     -√10              √9 √23
        -5         0         5
```

Remark. As you may suspect with this limited graphing experience, there are an infinite number of real numbers associated with the points on a line. The question probably arises in your mind whether the set of real numbers can be associated with *all* the points on the line, and whether the points on the line can be associated with *all* the real numbers. The answer is yes. There is a one-to-one correspondence between the real numbers and the points on a number line. Because the number line can be completely filled with the graphs of the real numbers, the set of real numbers is said to be **complete** or have the **property of completeness.** *There is a one-to-one correspondence between the elements of the set of real numbers and the points on a number line.* A more detailed discussion of the existence of irrational numbers and the notion of completeness is beyond our study here.

The various subsets of the set of real numbers are shown in the figure on page 218. You may wish to use this diagram as a reference as you complete the frames that follow.

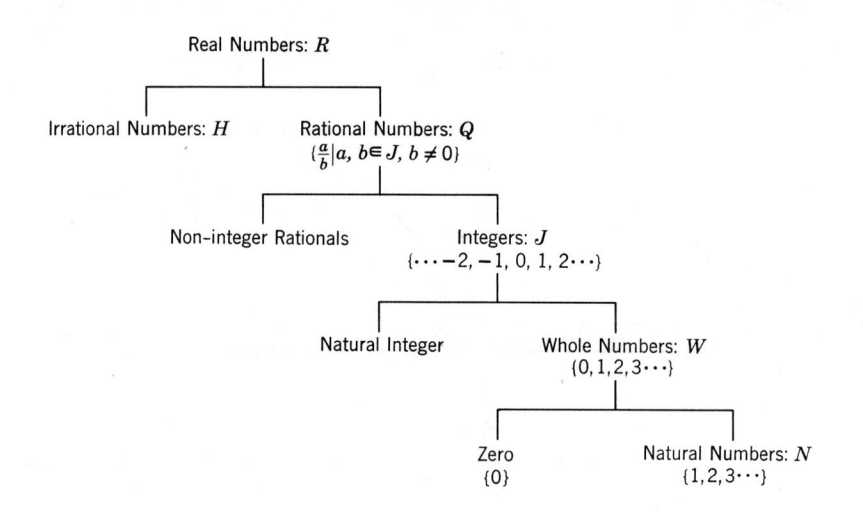

```
                    Real Numbers: R
         ┌─────────────────┴─────────────────┐
   Irrational Numbers: H           Rational Numbers: Q
                                   {a/b | a, b ∈ J, b ≠ 0}
                         ┌──────────────────┴──────────────┐
                  Non-integer Rationals              Integers: J
                                                {···−2, −1, 0, 1, 2···}
                                        ┌──────────────┴──────────────┐
                                   Natural Integer            Whole Numbers: W
                                                                {0, 1, 2, 3···}
                                                        ┌──────────┴──────────┐
                                                      Zero           Natural Numbers: N
                                                      {0}               {1, 2, 3···}
```

22. If $x \in H$, then x (is/is not) an element of Q.

is not (H and Q are disjoint.)

23. If $x \in Q$, then x (is/is not) an element of R.

is

24. If $x \in J$, then x (is/is not) an element of Q.

is

25. If $x \in \{$negative integers$\}$, then x (is/is not) an element of N.

is not

26. If $x \in N$, then x (is/is not) an element of Q.

is

27. $\{0\}$ (is/is not) a subset of J.

is

The Set of Real Numbers

28. W (is/is not) a subset of N.

    is not

29. {negative integers} (is/is not) a subset of Q.

    is (*All the integers are contained in Q.*)

30. J (is/is not) a subset of R.

    is

31. H (is/is not) a subset of R.

    is

32. Q (is/is not) a subset of R.

    is

33. $\{0\} \cup N =$ ___.

    W (Or $\{0, 1, 2, 3, \ldots\}$)

34. {negative integers} $\cup W =$ ___.

    J

35. {noninteger rationals} $\cup J =$ ___.

    Q

36. The set of real numbers has the property of completeness. The graphs of all of the real numbers completely "fill" the number line. Thus, $Q \cup H =$ ___.

    $Q \cup H = R$

220   UNIT V

37. The set of rational numbers and the set of irrational numbers are disjoint; that is, they have no elements in common. Thus $Q \cap H =$ _____ .

$\emptyset$

38. $J \cap W =$ _____ .

W (Or $\{0, 1, 2, 3, \ldots\}$)

39. $W \cap N =$ _____ .

N (Or $\{1, 2, 3, \ldots\}$)

40. $\{0\} \cap N =$ _____ .

$\emptyset$

41. $J \cap \{0\} =$ _____ .

$\{0\}$

42. $J \cap Q =$ _____ .

J

43. $W \cap R =$ _____ .

W

Remark. Some infinite sets of points determine line segments. Hence, as you will see in the following frames, the graphs of some infinite sets of real numbers are line segments.

44. The graph of $\{x | 0 < x < 4, x \in W\}$ consists of three points and is shown as

Graph $\{x | 3 < x < 7, x \in W\}$.

The Set of Real Numbers   221

45. The graph of $\{x|-2 < x < 4, x \in R\}$ is an infinite set of points and is shown as

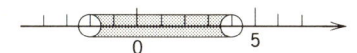

Open dots are used at −2 and 4 to indicate that these numbers are not elements in the given set. All the points between −2 and 4 are in the graph.
Graph $\{x|-3 < x < 2, x \in R\}$.

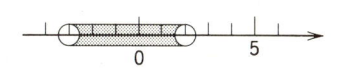

---

46. The graph of $\{x|1 \leq x < 4, x \in R\}$ is an infinite set of points and is shown as

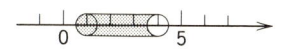

The graph of 1 is shown as a solid dot to indicate that 1 is an element of the given set. Graph $\{x|-1 < x \leq 3, x \in R\}$.

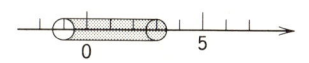

---

47. Graph $\{x| \ 1 \leq x < 5, x \in R\}$.

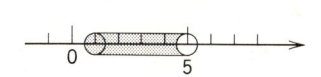

(*Graph includes points greater than or equal to* 1 *and less than* 5.)

48. Graph $\{x| \ -2 \leq x \leq 2, x \in R\}$.

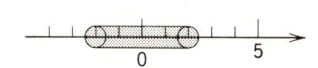

---

49. The graph of $\{x|x \geq 0, x \in R\}$ is an infinite set of points and is shown as

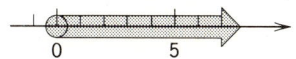

Graph $\{x|x < 0, x \in R\}$.

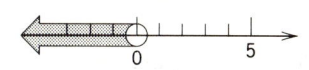

---

50. Graph $\{x | x \in R\}$.

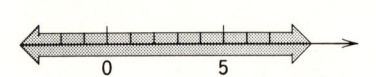

---

0 — 5 (number line graph)

Remark. The following frames will give you an opportunity to review what you have learned about the set of irrational numbers and the set of real numbers up to this point.

R51. The square root of a number is one of two equal factors of the number. The square roots of 81 are 9 and −9 because $(9)(9) = 81$ and $(\ \ )(\ \ ) = 81$.

$(-9)(-9) = 81$

R52. $\sqrt{81} = \underline{\ \ \ }$; $-\sqrt{81} = \underline{\ \ \ }$.

9; −9

R53. $\sqrt{\frac{25}{36}} = \underline{\ \ \ }$; $-\sqrt{\frac{9}{4}} = \underline{\ \ \ }$.

$\frac{5}{6}$; $-\frac{3}{2}$

R54. $\sqrt{b}$, $b \geq 0$, is the non negative number such that $\sqrt{b} \cdot \sqrt{b} = \underline{\ \ \ }$.

---

b

R55. A rational number can be expressed as the quotient of two integers. An irrational number (can/cannot) be expressed as the quotient of two integers.

---

cannot

R56. An irrational number can be approximated to any desired degree of accuracy by a _____ number.

---

rational

R57. Using the table on page 241, we have that $\sqrt{12} \approx \underline{\ \ \ }$ and $\sqrt{17} \approx \underline{\ \ \ }$.

3.464; 4.123

The Set of Real Numbers

R58. The set of real numbers are in a one-to-one correspondence with (all/some) of the points on a number line.

---
all

R59. The set $H = \{\text{irrational numbers}\}$ (is/is not) a subset of $R = \{\text{real numbers}\}$. The set $Q = \{\text{rational numbers}\}$ (is/is not) a subset of R.

---
is; is

R60. The set H (is/is not) a subset of the set Q.

---
is not

R61. $Q \cup H =$ ____, $Q \cap H =$ ____.

R; $\emptyset$

R62. Graph $\{x \mid 1 < x \leq 4, x \in R\}$.

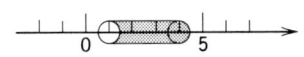

Remark. In Units II-IV of this program, we adopted some axioms for equality and order for the sets of numbers that we had been considering. In addition, we adopted eleven axioms for operations for the set of rational numbers which are the building blocks of that system. We shall also adopt these axioms for the system of real numbers.

Panel IV on pages 225 and 226 summarizes the axioms for the real numbers and the properties which are logical consequences of these axioms that have been introduced in this program. You may wish to use this panel as a reference to complete the frames that follow.

Panel IV

Axioms and properties for the set of real numbers

$$a,b,c,d \in R$$

EQUALITY AXIOMS

1. $a = a$.   Reflexive law of equality
2. If $a = b$, then $b = a$.   Symmetric law of equality
3. If $a = b$ and $b = c$, then $a = c$.   Transitive law of equality
4. If $a = b$, then a may be replaced by b or b may be replaced by a in any statement without affecting the truth or falsity of the statement.   Substitution law

ORDER AXIOMS

1. If $a < b$ and $b < c$, then $a < c$.   Transitive law of inequality
2. Exactly one of the following statements is true:
$$a < b, \quad a = b, \quad b < a.$$
     Trichotomy law
3. If $a, b > 0$, then
$$a + b > 0 \text{ and } a \cdot b > 0.$$

COMPLETENESS AXIOM

Every real number can be represented as a point on a number line and every point on a number line has a coordinate which is a real number.

AXIOMS FOR OPERATIONS

1. $a + b \in R$ — Closure law for addition
2. $a + b = b + a$ — Commutative law of addition
3. $(a + b) + c = a + (b + c)$ — Associative law of addition
4. $a + 0 = a$ and $0 + a = a$ — Identity law of addition
5. $a + (-a) = 0$ and $-a + a = 0$ — Additive inverse law
6. $a \cdot b \in R$ — Closure law for multiplication
7. $a \cdot b = b \cdot a$ — Commutative law of multiplication
8. $(a \cdot b) \cdot c = a \cdot (b \cdot c)$ — Associative law of multiplication
9. $a \cdot 1 = a$ and $1 \cdot a = a$ — Identity law of multiplication
10. $a \cdot \dfrac{1}{a} = 1$ and $\dfrac{1}{a} \cdot a = 1$ $(a \neq 0)$ — Multiplicative inverse law
11. $a \cdot (b + c) = a \cdot b + a \cdot c$ — Distributive law

PROPERTIES

1. If $a = b$, then $a + c = b + c$.
2. If $a = b$, then $a \cdot c = b \cdot c$.
3. $a \cdot 0 = 0$
4. $-a + (-b) = -(a + b)$
5. If $a > b > 0$, then $a + (-b) = a - b$; if $b > a > 0$, then $a + (-b) = -(b - a)$
6. $a - b = a + (-b)$
7. $a \cdot (-b) = -(a \cdot b)$
8. $(-a) \cdot (-b) = a \cdot b$
9. $\dfrac{a}{b} = a \cdot \dfrac{1}{b}$ $(b \neq 0)$
10. If $\dfrac{a}{b} = \dfrac{c}{d}$, then $a \cdot d = b \cdot c$; if $a \cdot d = b \cdot c$, then $\dfrac{a}{b} = \dfrac{c}{d}$ $(b, d \neq 0)$
11. $\dfrac{a \cdot c}{b \cdot c} = \dfrac{a}{b}$ $(b, c \neq 0)$
12. $\dfrac{a}{c} + \dfrac{b}{c} = \dfrac{a + b}{c}$ $(c \neq 0)$
13. $\dfrac{a}{c} - \dfrac{b}{c} = \dfrac{a - b}{c}$ $(c \neq 0)$
14. $\dfrac{a}{b} \cdot \dfrac{c}{d} = \dfrac{a \cdot c}{b \cdot d}$ $(b, d \neq 0)$
15. $\dfrac{1}{\frac{a}{b}} = \dfrac{b}{a}$ $(a, b \neq 0)$
16. $\dfrac{a}{b} \div \dfrac{c}{d} = \dfrac{a}{b} \cdot \dfrac{d}{c}$ $(b, c, d \neq 0)$

The Set of Real Numbers

Remark. In Unit IV of this program we established certain theorems that enabled us to rewrite sums, differences, products, and quotients of rational numbers.

Based on the way that we defined $\sqrt{b}$ ($b \geq 0$), we could also establish some theorems to rewrite sums, differences, products, and quotients of irrational numbers. However, this is beyond the scope of our objectives in this program.

The following frames will give you an opportunity to review some properties of the set of real numbers. You may want to use Panel IV as a reference to help you make the appropriate responses when you first start the sequence of frames.

In each of the following frames, the variables represent real numbers.

R63. The Reflexive law of equality states that $a = \underline{}$.

$a = a$

R64. If $a = 7$, then the statement that $7 = a$ is justified by the _____ law of equality.

Symmetric

R65. If $r = s$ and $s = 3$, then by the Transitive law of equality, or the Substitution law, we may write $r = \underline{}$.

$r = 3$

R66. If $4 + a = x$ and $a = r$, then by the _____ law, we may write $4 + r = x$.

Substitution

R67. The Transitive law of inequality states that if $a < b$ and $b < c$, then $a \underline{} c$.

$a < c$

R68. By the Trichotomy law exactly one of the following statements is true. $a < b$, $a = b$, or $b \underline{} a$.

$b < a$

228   UNIT V

R69.   If $b > 0$, then $6 + b$ ____ 0 and $6 \cdot b$ ____ 0.

       $6 + b > 0$; $6 \cdot b > 0$

R70.   The property of completeness states that every real number can be associated with exactly one _____ on a number line and every point on a number line can be associated with exactly one _____ number.
       ----
       point; real

R71.   If $a \in R$, then the statement that $a + 9$ is also in R is justified by the _____ law for addition.

       Closure

R72.   In the statement $8 + 3 = 3 + 8$, the (grouping/order) of the numbers has been changed. This statement is justified by the _____ law of addition.
       ----
       order; Commutative

R73.   In the statement $(a + 9) + c = a + (9 + c)$, the (grouping/order) of the numbers has been changed. This statement is justified by the _____ law of addition.
       ----
       grouping; Associative

R74.   The number 0 is the identity element for _____.
       ----
       addition

R75.   The Identity law of addition states that for all a, $a + 0 =$ ____.

       $a + 0 = a$

R76.   The Additive inverse law states that the sum of a and the _____ of a equals 0.
       ----
       additive inverse   (*Or negative*)

The Set of Real Numbers

R77. Symbolically, $a + (\quad) = 0$ and $\quad + a = 0$.

$a + (-a) = 0;\ -a + a = 0$

R78. If $x, y \in R$, then the statement that the product $x \cdot y$ is also in R is justified by the _____ law for multiplication.

Closure

R79. In the statement $7 \cdot 8 = 8 \cdot 7$, the (grouping/order) of the numbers has been changed. This statement is justified by the _____ law of multiplication.

---
order; Commutative

R80. In the statement $(8 \cdot 6) \cdot 9 = 8 \cdot (6 \cdot 9)$, the (grouping/order) of the numbers has been changed. This statement is justified by the _____ law of multiplication.

---
grouping; Associative

R81. The Identity law of multiplication states that for all a, $a \cdot 1 = \_\_\_$.

$a \cdot 1 = a$

R82. The Multiplicative inverse law states that the product of a $(a \neq 0)$ and the _____ _____ of a equals 1. Symbolically, $a \cdot \_\_\_ = \_\_\_$.

---
multiplicative inverse; $a \cdot \dfrac{1}{a} = 1$

R83. The statement $7 \cdot (a + b) = 7 \cdot a + 7 \cdot b$ is justified by the _____ law.

Distributive

R84. The statement "If $a = b$, then $a + 9 = \_\_\_$" is justified by the Addition law of equality.

$a + 9 = b + 9$

R85. The statement, "If $a = b$, then $7 \cdot a = $ _____" is justified by the Multiplication law of equality.

$7 \cdot a = 7 \cdot b$

R86. For all a, $a \cdot 0 = $ _____ .

$a \cdot 0 = 0$

R87. The sum of two positive real numbers is (always/sometimes/never) a positive real number.

---

always

R88. The sum of two negative real numbers is (always/sometimes/never) a negative real number.

---

always

R89. Symbolically, $-a + (-b) = -($ _____ $)$.

$-(a + b)$

R90. $^{+}2 + {}^{+}13 = $ _____ and $^{-}2 + {}^{-}13 = $ _____ .

$^{+}15$; $^{-}15$

R91. The sum of a positive real number and a negative real number is (always/sometimes/never) a positive real number.

---

sometimes

R92. $^{-}3 + {}^{+}11 = $ _____ and $^{+}3 + {}^{-}11 = $ _____ .

$^{+}8$; $^{-}8$

R93. $5 + 7 = $ _____ ; $-5 + (-7) = $ _____ ; $5 + (-7) = $ _____ ; $-5 + 7 = $ _____ .

12; −12; −2; 2

The Set of Real Numbers

R94. The difference $a - b$ can be expressed as the sum $a + (\underline{})$.

$a + (-b)$

R95. $7 - 2 = 7 + (-2) = \underline{}$ and $2 - 7 = 2 + (-7) = \underline{}$.

5; −5

R96. The product of two positive real numbers is (always/sometimes/never) a positive real number.

---
always

R97. The product of a positive real number and a negative real number is (always/sometimes/never) a negative real number.

---
always

R98. The product $a \cdot (-b) = -(\underline{})$.

$-(a \cdot b)$

R99. The product of two negative real numbers is (always/sometimes/never) a positive real number.

---
always

R100. The product $(-a) \cdot (-b) = \underline{}$.

$a \cdot b$

R101. $3 \cdot 11 = \underline{}$; $8 \cdot (-9) = \underline{}$; $(-5)(-6) = \underline{}$.

33; −72; 30

R102. If $b \neq 0$, then $\dfrac{a}{b} = a \cdot (\underline{})$.

$a \cdot \left(\dfrac{1}{b}\right)$

UNIT V

R103. If $b,d \neq 0$, then $\dfrac{a}{b} = \dfrac{c}{d}$ if $a \cdot d =$ _____ .

$a \cdot d = b \cdot c$

R104. $\dfrac{2}{5} = \dfrac{6}{15}$ because _____ $\cdot$ _____ = _____ $\cdot$ _____ .

$2 \cdot 15 = 5 \cdot 6$

R105. If $b,c \neq 0$, then $\dfrac{a \cdot c}{b \cdot c} =$ _____ .

$\dfrac{a \cdot c}{b \cdot c} = \dfrac{a}{b}$

R106. The numerator and the denominator of the fraction $\dfrac{35}{56}$ have the common factor, _____ ; hence $\dfrac{35}{56} = \dfrac{}{8}$.

$7; \dfrac{5}{8}$

R107. If $c \neq 0$, then $\dfrac{a}{c} + \dfrac{b}{c} =$ _____ .

$\dfrac{a + b}{c}$

R108. $\dfrac{1}{8} + \dfrac{10}{8} =$ _____ ; $\dfrac{2}{3} + \dfrac{5}{7} = \dfrac{2 \cdot 7}{3 \cdot 7} + \dfrac{5 \cdot 3}{7 \cdot 3} =$ _____ .

$\dfrac{11}{8}; \dfrac{29}{21}$

R109. If $c \neq 0$, then $\dfrac{a}{c} - \dfrac{b}{c} =$ _____ .

$\dfrac{a - b}{c}$

The Set of Real Numbers

R110. $\dfrac{8}{11} - \dfrac{3}{11} =$ _____ ; $\dfrac{3}{5} - \dfrac{1}{2} = \dfrac{3 \cdot 2}{5 \cdot 2} - \dfrac{1 \cdot 5}{2 \cdot 5} =$ _____ .

$\dfrac{5}{11}$; $\dfrac{1}{10}$

R111. If $b, d \neq 0$, then $\dfrac{a}{b} \cdot \dfrac{c}{d} =$ _____ .

$\dfrac{a \cdot c}{b \cdot d}$

R112. $\dfrac{2}{5} \cdot \dfrac{3}{7} = \dfrac{2 \cdot 3}{5 \cdot 7} =$ _____ ; $\dfrac{3}{7} \cdot \dfrac{1}{2} =$ _____ .

$\dfrac{2 \cdot 3}{5 \cdot 7} = \dfrac{6}{35}$; $\dfrac{3}{14}$

R113. If $b, c, d \neq 0$, then $\dfrac{a}{b} \div \dfrac{c}{d} =$ _____ $\cdot$ _____ .

$\dfrac{a}{b} \cdot \dfrac{d}{c}$

R114. $\dfrac{1}{7} \div \dfrac{3}{4} = \dfrac{1}{7} \cdot$ _____ = _____ ; $\dfrac{2}{3} \div \dfrac{1}{2} =$ _____ .

$\dfrac{1}{7} \cdot \dfrac{4}{3} = \dfrac{4}{21}$; $\dfrac{4}{3}$

R115. If $a, b \neq 0$, then $\dfrac{1}{\dfrac{a}{b}} =$ _____ .

$\dfrac{b}{a}$

R116. $\dfrac{\dfrac{1}{4}}{\dfrac{1}{5}} =$ _____ ; $\dfrac{1}{\dfrac{1}{3}} =$ _____ .

$\dfrac{5}{4}$; 3

R117. The set of rational numbers $\{\frac{a}{b} \mid a,b \in J, b \neq 0\}$ can be associated with a(n) (finite/infinite) number of points on a number line.

infinite

R118. The graphs of the rational numbers (do/do not) completely fill the number line.

do not

R119. Numbers that can be associated with points on the number line and that are not rational numbers are called _____ numbers.

irrational

R120. The union of the set of rational numbers and the set of irrational numbers is called the set of _____ numbers.

real

R121. There is a _____ _____ _____ correspondence between the elements of the set of real numbers and the set of points on a number line.

one-to-one

R122. Because there is a one-to-one correspondence between the elements of the set of real numbers and the points on a number line, the set of real numbers is said to have the property of _____.

completeness

Remark. This remark concludes the program. You have now been introduced to several number systems all of which have been based on three things: a set of elements, at least one well-defined operation on these elements, and a set of axioms that govern the behavior of the elements under these operations (see Axioms for Operations, Panel IV). The remainder of each of these systems (its structure) consists of the logical consequences (theorems) of these basic building blocks. Primitive notions such as closure, commutativity, associativity, and so on, are associated with the set of whole numbers and these same ideas are embodied into each successive extension of this number system.

▼

The Set of Real Numbers

Try the self-evaluation test that follows. Score your paper from the answers given on page 279.

The bibliography on page 240 provides a list of textbooks that are particularly suited for further study of the set of real numbers and its subsets that have been introduced in this program.

UNIT V
Self-Evaluation Test

1. Each of two equal factors of a positive real number is called the _____ _____ of the number.
2. $\sqrt{b}$, $(b \geq 0)$ is the non negative number such that $\sqrt{b} \cdot \sqrt{b} =$ _____.
3. $\sqrt{36} =$ _____.
4. $\sqrt{7} \cdot \sqrt{7} =$ _____.
5. Real numbers which cannot be expressed in the form $\frac{a}{b}$, where $a, b \in J$, are called _____ numbers.
6. Graph $\{-\sqrt{36}, -\sqrt{5}, \sqrt{34}\}$.
7. $Q \cup H =$ _____; $Q \cap H =$ _____.
8. The graphs of the set of real numbers completely fill the number line. Hence, the set is said to have the property of _____.
9. Graph $\{x | -3 < x < 1, x \in R\}$.
10. The statement $5 \cdot (x + y) = 5 \cdot x + 5 \cdot y$ is justified by the _____ law.
11. The sum of two negative real numbers is (always/sometimes/never) a negative real number.
12. The sum of a positive real number and a negative real number is (always/sometimes/never) a negative real number.
13. The difference $a - b$ can be expressed as the sum _____.
14. The product of a positive real number and a negative real number is (always/sometimes/never) a negative real number.
15. The product of two negative real numbers is (always/sometimes/never) a positive real number.
16. The quotient $\frac{a}{b}$ can be expressed as the product _____.

236

The Set of Real Numbers

17. The quotient $\frac{a}{b} \div \frac{c}{d}$ can be expressed as the product _____ .

Consider the set R of real numbers and the following subsets of R:
$N = \{\text{natural numbers}\}$, $W = \{\text{whole numbers}\}$, $J = \{\text{integers}\}$, $Q = \{\text{rational numbers}\}$, and $H = \{\text{irrational numbers}\}$.

18. Which of the above sets *do not* contain -3 as a member?

19. Which of the above sets are not closed for subtraction?

20. Which of the above sets are not closed for division?

Appendixes

Bibliography

1. Allendoerfer, C. B., *Principles of Arithmetic and Geometry for Elementary School Teachers*, The Macmillan Company, New York, 1971
2. Brumfiel and Krause, *Elementary Mathematics for Teachers*, Addison-Wesley Publishing Co., Inc., Reading, Mass., 1969
3. Fehr and Phillips, *Teaching Modern Mathematics in the Elementary School*, Addison-Wesley Publishing Co. Inc., Reading, Mass., 1967
4. Forbes and Eicholz, *Mathematics for Elementary Teachers*, Addison-Wesley Publishing Co. Inc., Reading, Mass., 1971
5. Graham, M., *Modern Elementary Mathematics*, Harcourt, Brace, & World, Inc., New York, 1970
6. Hutton, Rex L., *Number Systems–An Intuitive Approach*, Educational Publishers, Scranton, Pa., 1971
7. Kelley and Richert, *Elementary Mathematics for Teachers*, Holden Day, San Francisco, 1970
8. Nichols, E. D. and Swain, R. L., *Mathematics for the Elementary School Teacher*, Holt, Rinehart & Winston, New York, 1971
9. Peterson and Hashisaki, *Theory of Arithmetic*, John Wiley & Sons, Inc., New York, Third Edition, 1971
10. Scandura, J. M., *Mathematics: Concrete Behavioral Foundations*, Harper & Row, New York, 1971
11. Smith, S. E., *Explorations in Elementary Mathematics*, Prentice-Hall, Englewood Cliffs, New Jersey, 1971
12. Spector, L., *Liberal Arts Mathematics*, Addison-Wesley, Reading, Mass., 1971
13. Willerding, M. F., *Elementary Mathematics: Its Structure and Concepts*, John Wiley & Sons, Inc., New York, Second Edition, 1970
14. Thirty-first yearbook, *Historical Topics for the Mathematics Classroom*, NCTM, Washington, D.C., 1969

Table of Square Roots

No.	Sq. Root	No.	Sq. Root	No.	Sq. Root	No.	Sq. Root	No.	Sq. Root
1	1.000	21	4.583	41	6.403	61	7.810	81	9.000
2	1.414	22	4.690	42	6.481	62	7.874	82	9.055
3	1.732	23	4.796	43	6.557	63	7.937	83	9.110
4	2.000	24	4.899	44	6.633	64	8.000	84	9.165
5	2.236	25	5.000	45	6.708	65	8.062	85	9.220
6	2.449	26	5.099	46	6.782	66	8.124	86	9.274
7	2.646	27	5.196	47	6.856	67	8.185	87	9.327
8	2.828	28	5.292	48	6.928	68	8.246	88	9.381
9	3.000	29	5.385	49	7.000	69	8.307	89	9.434
10	3.162	30	5.477	50	7.071	70	8.367	90	9.487
11	3.317	31	5.568	51	7.141	71	8.426	91	9.539
12	3.464	32	5.657	52	7.211	72	8.485	92	9.592
13	3.606	33	5.745	53	7.280	73	8.544	93	9.644
14	3.742	34	5.831	54	7.348	74	8.602	94	9.695
15	3.873	35	5.916	55	7.416	75	8.660	95	9.747
16	4.000	36	6.000	56	7.483	76	8.718	96	9.798
17	4.123	37	6.083	57	7.550	77	8.775	97	9.849
18	4.243	38	6.164	58	7.616	78	8.832	98	9.899
19	4.359	39	6.245	59	7.681	79	8.888	99	9.950
20	4.472	40	6.325	60	7.746	80	8.944	100	10.000

Exercise Sets

Exercise Set Ia

1. Identify each symbol.
    a. { }   b. =   c. ≠   d. ∈
    e. ∉   f. ∅   g. ⊂

2. Let U = {letters of the English alphabet} and A = {vowels in the English alphabet}. Replace the comma in the following pairs with either ∈ or ∉.
    a. r, A b. r, U c. i, A
    d. b, A e. $m, \{m,n\}$ f. $\{m\}, \{m,n\}$
    g. e, U h. g, A i. g, U

3. Let $K = \{a, b, c\}$, $L = \{c, a, b\}$, $M = \{b, c\}$, $N = \{a, b\}$, and $P = \{a\}$. Replace the comma in the following pairs with either = or ≠.
    a. K, L b. M, N c. a, P
    d. $\{a\}, P$ e. $K, 3$ f. $n(K), 3$
    g. $M, 2$ h. $n(M), 2$ i. $n(P), 1$

4. Let $K = \{a, b, c\}$, $L = \{c, a, b\}$, $M = \{b, c\}$, $N = \{a, b\}$, and $P = \{a\}$. Replace the comma in the following pairs with ∈ or ⊂.
    a. b, K b. $\{c\}, L$ c. $\emptyset, N$
    d. $\emptyset, P$ e. P, N f. M, L
    g. P, P h. b, M i. a, N

5. Let $B = \{e, f, g\}$. List the subsets of B that contain:
    a. three members   b. two members
    c. one member   d. no members

242

Appendixes

6. Let $C = \{k, l, m, n\}$. List the subsets of C that contain:

    a. four members   b. three members   c. two members

    d. one member   e. no members

Let $U = \{\text{letters of the English alphabet}\}$, $A = \{a, b, c, d\}$, $B = \{c, d\}$, $C = \{c, d, e\}$, $D = \{e, f, g, h\}$, and $E = \{b, c, d, e\}$.

7. Which of the following are true and which are false?

    a. D and E are disjoint   b. A and D are disjoint

    c. A is equivalent to B d. A is equivalent to E

    e. $D = E$ f. $A \neq C$ g. $A \subset U$

    h. $A \subset B$ i. $C \subset B$

8. Which of the following are true and which are false?

    a. $C \subset E$ b. $D \subset U$ c. $B \subset D$

    d. $\emptyset \subset B$ e. $\emptyset \subset E$ f. $e \in A$

    g. $e \subset C$ h. $e \notin U$ i. $j \notin U$

Answers Ia

1. a. Braces   b. is equal to   c. is not equal to   d. is an element of

   e. is not an element of   f. null set   g. is a subset of

2. a. $\notin$ b. $\in$ c. $\in$ d. $\notin$ e. $\in$

   f. $\notin$ g. $\in$ h. $\notin$ i. $\in$

3. a. $=$ b. $\neq$ c. $\neq$ d. $=$ e. $\neq$

   f. $=$ g. $\neq$ h. $=$ i. $=$

4. a. $\in$ b. $\subset$ c. $\subset$ d. $\subset$ e. $\subset$

   f. $\subset$ g. $\subset$ h. $\in$ i. $\in$

5. a. $\{e, f, g\}$ b. $\{e, f\}, \{e, g\}, \{f, g\}$ c. $\{e\}, \{f\}, \{g\}$ d. $\emptyset$

6. a. $\{k,l,m,n\}$
   b. $\{k,l,m\}$, $\{k,l,n\}$, $\{k,m,n\}$, $\{l,m,n\}$
   c. $\{k,l\}$, $\{k,m\}$, $\{k,n\}$, $\{l,m\}$, $\{l,n\}$, $\{m,n\}$
   d. $\{k\}$, $\{l\}$, $\{m\}$, $\{n\}$
   e. $\emptyset$

7. a. false b. true c. false d. true e. false
   f. true g. true h. false i. false

8. a. true b. true c. false d. true e. true
   f. false g. false h. false i. false

Appendixes

Exercise Set Ib

1. Identify each symbol.

   a. ∪   b. ∩   c. U   d. A' e. (a,b) f. $A \times B$

2. Let $A = \{j,k,l\}$, $B = \{k,l,m\}$, $C = \{j,k\}$, $D = \{l,m\}$, and $E = \{k\}$.
   List the elements in each of the following sets.

   a. $A \cup B$
   b. $B \cup C$
   c. $C \cup D$
   d. $D \cup E$
   e. $A \cup C$
   f. $C \cup E$
   g. $B \cap D$
   h. $D \cap C$
   i. $B \cap C$

3. Let $U = \{d,e,f,g,h\}$, $A = \{d,e\}$, $B = \{e,f\}$, $C = \{g,h\}$, and $D = \{h\}$.
   List the elements in each of the following sets.

   a. A'
   b. B'
   c. C'
   d. D'
   e. $A' \cup B$
   f. $A \cup B'$
   g. $B' \cup C$
   h. $B \cup C'$
   i. $C' \cap D$

4. Let A be a finite subset of U. Complete each of the following.

   a. $A \cup A =$ _____
   b. $A \cap A =$ _____
   c. $A \cup \emptyset =$ _____
   d. $A \cup A' =$ _____
   e. $A \cap A' =$ _____
   f. $A \cup U =$ _____
   g. $A \cap U =$ _____
   h. $A' \cup \emptyset =$ _____
   i. $A' \cap \emptyset =$ _____

5. Write the set illustrated by the shaded region of each Venn diagram.

   a. [Venn diagram with U, R, S]
   b. [Venn diagram with U, K, L]
   c. [Venn diagram with U, A, B, C]

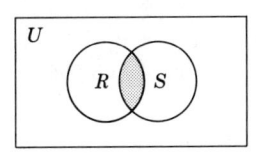

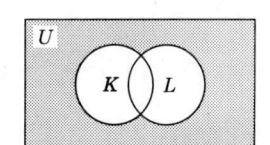

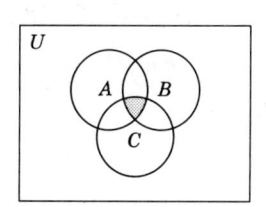

6. Let $A = \{a,e\}$, $B = \{e\}$, $C = \{d,e,f\}$, and $D = \{c,h\}$. Write the Cartesian products:

a. $A \times B$
b. $B \times A$
c. $A \times C$
d. $C \times A$
e. $A \times D$
f. $D \times A$
g. $B \times C$
h. $C \times B$
i. $B \times D$

Answers Ib

1. a. The union of   b. the intersection of   c. the universal set
   d. the complement of set A e. the ordered pair of numbers whose first component is a and whose second component is b
   f. the Cartesian product of sets A and B

2. a. $\{j,k,l,m\}$ b. $\{j,k,l,m\}$ c. $\{j,k,l,m\}$
   d. $\{k,l,m\}$ e. $\{j,k,l\}$ f. $\{j,k\}$
   g. $\{l,m\}$ h. $\emptyset$ i. $\{k\}$

3. a. $\{f,g,h\}$ b. $\{d,g,h\}$ c. $\{d,e,f\}$
   d. $\{d,e,f,g\}$ e. $\{e,f,g,h\}$ f. $\{d,e,g,h\}$
   g. $\{d,g,h\}$ h. $\{d,e,f\}$ i. $\emptyset$

4. a. A b. A c. A
   d. U e. $\emptyset$ f. U
   g. A h. A' i. $\emptyset$

5. a. $R \cap S$ b. $(K \cup L)'$ c. $A \cap B \cap C$

6. a. $\{(a,e), (e,e)\}$
   b. $\{(e,a), (e,e)\}$
   c. $\{(a,d), (a,e), (a,f), (e,d), (e,e), (e,f)\}$
   d. $\{(d,a), (d,e), (e,a), (e,e), (f,a), (f,e)\}$
   e. $\{(a,c), (a,h), (e,c), (e,h)\}$
   f. $\{(c,a), (c,e), (h,a), (h,e)\}$
   g. $\{(e,d), (e,e), (e,f)\}$
   h. $\{(d,e), (e,e), (f,e)\}$
   i. $\{(e,c), (e,h)\}$

Exercise Set IIa

1. Identify each symbol.
   a. $<$
   b. $>$
   c. $\{a,b,c,\ldots\}$
   d. W
   e. N
   f. $\{a|a\ldots\}$

2. Replace the comma in the following pairs with the correct order symbols, $<$, $>$, or $=$.
   a. 5, 7
   b. 11, 3
   c. 7, 7
   d. 9, 8
   e. 6, 0
   f. 0, 5
   g. 6, 1
   h. 1, 6
   i. 6, 6

3. List the elements in the following sets.
   a. $\{a|a<5, a\in N\}$
   b. $\{a|a<7, a\in N\}$
   c. $\{b|b\leqslant 4, b\in W\}$
   d. $\{b|b\leqslant 2, b\in W\}$
   e. $\{a|a>7, a\in W\}$
   f. $\{a|a>5, a\in W\}$
   g. $\{a|5\leqslant a<7, a\in N\}$
   h. $\{b|3<b\leqslant 6, b\in N\}$

4. State which of the following sets are finite and which are infinite.
   a. $\{x|x<8, x\in N\}$
   b. $\{x|x>0, x\in W\}$
   c. $\{x|1<x<3, x\in W\}$
   d. $\{x|5\leqslant x<9, x\in N\}$

5. Write each set in set-builder notation. Use x as the variable. (There may be more than one way to write each set.)
   a. $\{4,5\}$
   b. $\{2,3,4\}$
   c. $\{0,1,2,\ldots\}$
   d. $\{3,4,5,6,\ldots\}$
   e. $\{5,6,7,8,9\}$
   f. $\{5,6,7,8,9,\ldots\}$

6. Graph each set.
   a. $\{0, 1\}$
   b. $\{1, 2\}$
   c. $\{3,4,5\}$
   d. $\{5, 6, 7\}$
   e. $\{7, 8, 9, \ldots\}$
   f. $\{4,5,6,\ldots\}$
   g. $\{x|x<4, x\in W\}$
   h. $\{x|x<4, x\in N\}$
   i. $\{x|x>3, x\in W\}$

Answers IIa

1. a. Is less than   b. is greater than   c. an infinite set whose first three elements are $a,b,$ and c d. the set of whole numbers   e. the set of natural numbers   f. the set of all a such that a is ...

2. a. $<$ b. $>$ c. $=$ d. $>$
   e. $>$ f. $<$ g. $>$ h. $<$ i. $=$

3. a. $\{1,2,3,4\}$ b. $\{1,2,3,4,5,6\}$ c. $\{0,1,2,3,4\}$
   d. $\{0,1,2\}$ e. $\{8,9,10,\ldots\}$ f. $\{6,7,8,\ldots\}$
   g. $\{5,6\}$ h. $\{4,5,6\}$

4. a. Finite   b. infinite   c. finite   d. finite

5. a. $\{x | 3 < x < 6, x \in W\}$ b. $\{x | 1 < x < 5, x \in W\}$
   c. $\{x | x \geqslant 0, x \in W\}$ d. $\{x | x \geqslant 3, x \in W\}$
   e. $\{x | 5 \leqslant x \leqslant 9, x \in W\}$ f. $\{x | x > 4, x \in W\}$

6. a.-i. [number line graphs]

Exercise Set IIb

Name the axiom that justifies each statement.

1. If $a = 4$ and $4 = b$, then $a = b$.
2. If $a = b$ and $5 < a$, then $5 < b$.
3. If $a < 2$ and $2 < b$, then $a < b$.
4. $a = a$.
5. If $a = 6$ and $b < a$, then $b < 6$.
6. If $a < b$ and $b < 7$, then $a < 7$.
7. If $a = 3$, then $3 = a$.
8. If $8 = b$, then $b = 8$.
9. If $a = 9$ and $b < 9$, then $b < a$.
10. If $a = 1$ and $1 = b$, then $a = b$.
11. If $a = b$ and $10 < a$, then $10 < b$.
12. If $a < 4$ and $4 < b$, then $a < b$.
13. $b = b$.
14. If $a = 5$, then $5 = a$.
15. $a < 2$, $a = 2$ or $a > 2$.
16. If $b = 6$ and $6 = a$, then $b = a$.
17. If $a = b$ and $b > 1$, then $a > 1$.
18. $b < 7$, $b = 7$, or $b > 7$.
19. If $a = 3$ and $a = b$, then $b = 3$.
20. If $a = 6$, then $6 = a$.

Answers IIb

1. Transitive law of equality
2. Substitution law
3. Transitive law of inequality
4. Reflexive law of equality
5. Substitution law
6. Transitive law of inequality
7. Symmetric law of equality
8. Symmetric law of equality
9. Substitution law
10. Transitive law of equality
11. Substitution law
12. Transitive law of inequality
13. Reflexive law of equality
14. Symmetric law of equality
15. Trichotomy law
16. Transitive law of equality
17. Substitution law
18. Trichotomy law
19. Substitution law
20. Symmetric law of equality

Exercise Set IIc

1. Let $A = \{2,4\}$, $B = \{3,5,7\}$, $C = \{6,8,10,12\}$, $D = \{1\}$, and $E = \{9\}$.
Write each of the following sets.

a. $A \cup B$
b. $A \cup C$
c. $A \cup D$
d. $A \cup E$
e. $B \cup C$
f. $B \cup D$
g. $B \cup E$
h. $C \cup D$
i. $C \cup E$

2. Let $E = \{2,4,6,8\}$, $G = \{1,3,5\}$, $H = \{7,9\}$, and $I = \{10\}$.
How many elements are there in each of the following sets?

a. E
b. G
c. H
d. I
e. $E \cup G$
f. $E \cup H$
g. $E \cup I$
h. $G \cup H$
i. $G \cup I$

3. Name the axiom to justify each statement. $a,b,c \in W$.

a. $3 + (a + b) = (a + b) + 3$
b. $5 + 0 = 5$
c. $(5 + 2) + 1 = 5 + (2 + 1)$
d. $a + (b + c) \in W$
e. $3 + (4 + c) = 3 + (c + 4)$
f. $(4 + b) + (1 + c) = (1 + c) + (4 + b)$
g. If $a + b = 7$ and $b = 4$, then $a + 4 = 7$
h. $(2 + b) + (3 + c) = [(2 + b) + 3] + c$
i. If $(a + b) + c = 9$ and $a = 3$ and $c = 4$, then $(3 + b) + 4 = 9$

4. Write each of the following as a basic numeral.

a. $5 + 7$
b. $8 + 9$
c. $(9 + 2) + 5$
d. $6 + (9 + 3)$
e. $6 + (18 + 4)$
f. $12 + 20 + 8$
g. $6 + 12 + 14$
h. $4 + 0 + 12$
i. $18 + 0 + 6$

5. Complete the following.

a. If $a = b$, then $a + 2 =$ _____.
b. If $a = 5$, then $a + 7 =$ _____.
c. If $3 = b$, then $3 + 11 =$ _____.
d. If $b = 4$, then _____ $= 4 + 9$.

Appendixes

6. Which of the following sets are closed for addition? If any are not closed, give an example to justify your conclusion.

a. $\{0\}$ b. $\{1,2\}$ c. $\{1,3,5,7\}$
d. $\{4,5,6,\ldots\}$ e. $\{2,4,6,\ldots\}$ f. $\{a|a > 9, a \in W\}$

Answers IIc

1. a. $\{2,3,4,5,7\}$ b. $\{2,4,6,8,10,12\}$ c. $\{1,2,4\}$
   d. $\{2,4,9\}$ e. $\{3,5,6,7,8,10,12\}$ f. $\{1,3,5,7\}$
   g. $\{3,5,7,9\}$ h. $\{1,6,8,10,12\}$ i. $\{6,8,9,10,12\}$

2. a. 4 b. 3 c. 2 d. 1 e. 7
   f. 6 g. 5 h. 5 i. 4

3. a. Commutative law of addition b. Identity law of addition
   c. Associative law of addition d. Closure law for addition
   e. Commutative law of addition f. Commutative law of addition
   g. Substitution law h. Associative law of addition
   i. Substitution law

4. a. 12 b. 17 c. 16 d. 18 e. 28
   f. 40 g. 32 h. 16 i. 24

5. a. $b + 2$ b. $5 + 7$ c. $b + 11$ d. $b + 9$

6. a. Closed for addition
   b. Not closed for addition; 1 + 2 is not an element in the set
   c. Not closed for addition; 1 + 3 is not an element in the set
   d. Closed for addition
   e. Closed for addition
   f. Closed for addition

Exercise Set IId

1. Let $A = \{1,3\}$, $B = \{2,4,6\}$, $C = \{3,5,7,9\}$, and $D = \{1\}$. Write each of the following sets of ordered pairs.

   a. $A \times B$
   b. $A \times C$
   c. $A \times D$
   d. $B \times C$
   e. $B \times D$
   f. $C \times D$
   g. $B \times A$
   h. $D \times A$
   i. $D \times C$

2. Let $G = \{1,3,5,7\}$, $H = \{2,4,6\}$, $I = \{1,2\}$, and $J = \{8\}$. How many elements are there in each of the following sets?

   a. G
   b. H
   c. I
   d. J
   e. $H \times J$
   f. $I \times J$
   g. $H \times G$
   h. $I \times G$
   i. $J \times G$

3. Name the axiom used to justify each statement. $a,b,c \in W$.

   a. $1 \cdot 53 = 53$
   b. $25 \cdot a \in W$
   c. $21 \cdot 4 = 4 \cdot 21$
   d. $(4 \cdot a) \cdot 2 = 4 \cdot (a \cdot 2)$
   e. $a + b = 1 \cdot (a + b)$
   f. $5 \cdot b \cdot (6 + a) = (6 + a) \cdot 5 \cdot b$
   g. $7 \cdot a \cdot b \in W$
   h. $(4 \cdot a)(3 \cdot b) = (4 \cdot a \cdot 3) \cdot b$

4. Write each of the following as a basic numeral.

   a. $3 \cdot 8$
   b. $4 \cdot 9$
   c. $(5 \cdot 6) \cdot 2$
   d. $2 \cdot (4 \cdot 3)$
   e. $2 \cdot 6 \cdot 4$
   f. $5 \cdot 4 \cdot 3$
   g. $6 \cdot 2 \cdot 1$
   h. $8 \cdot 1 \cdot 1$
   i. $13 \cdot 1 \cdot 2$

5. Complete the following.

   a. If $a = b$, then $7 \cdot a =$ _____.
   b. If $a = 5$, then $3 \cdot a =$ _____.
   c. If $4 = b$, then $6 \cdot 4 =$ _____.
   d. If $b = 7$, then _____ $= 2 \cdot 7$.
   e. $4 \cdot 0 =$ _____.
   f. $0 \cdot 4 \cdot 6 =$ _____.

Appendixes

6. Use the distributive law to write each of the following products as the sum of two products.

   a. 2(10 + 4)   b. 4(10 + 5)   c. 6(3 + 9)
   d. 5(8 + 6)   e. (10 + 6)3   f. (10 + 9)7
   g. (20 + 5)6   h. (30 + 7)8   i. (40 + 6)5

7. Which of the following sets are closed for multiplication? If any are not closed, give an example to justify your conclusion.

   a. $\{0\}$ b. $\{1\}$ c. $\{0,1\}$
   d. $\{1,2,3\}$ e. $\{4,5,6,7,\ldots\}$ f. $\{a|a > 2, a \in W\}$

Answers IId

1. a. $\{(1,2),(1,4),(1,6),(3,2),(3,4),(3,6)\}$
   b. $\{(1,3),(1,5),(1,7),(1,9),(3,3),(3,5),(3,7),(3,9)\}$ c. $\{(1,1),(3,1)\}$
   d. $\{(2,3),(2,5),(2,7),(2,9),(4,3),(4,5),(4,7),(4,9),(6,3),(6,5),(6,7),(6,9)\}$
   e. $\{(2,1),(4,1),(6,1)\}$ f. $\{(3,1),(5,1),(7,1),(9,1)\}$
   g. $\{(2,1),(2,3),(4,1),(4,3),(6,1),(6,3)\}$ h. $\{(1,1),(1,3)\}$
   i. $\{(1,3),(1,5),(1,7),(1,9)\}$

2. a. 4   b. 3   c. 2   d. 1   e. 3   f. 2   g. 12   h. 8   i. 4

3. a. Identity law of multiplication   b. Closure law for multiplication
   c. Commutative law of multiplication   d. Associative law of multiplication
   e. Identity law of multiplication   f. Commutative law of multiplication
   g. Closure law for multiplication   h. Associative law of multiplication

4. a. 24   b. 36   c. 60   d. 24   e. 48   f. 60   g. 12   h. 8   i. 26
5. a. $7 \cdot b$ b. $3 \cdot 5$ c. $6 \cdot b$ d. $2 \cdot b$ e. 0   f. 0

6. a. $2 \cdot 10 + 2 \cdot 4$ b. $4 \cdot 10 + 4 \cdot 5$ c. $6 \cdot 3 + 6 \cdot 9$
   d. $5 \cdot 8 + 5 \cdot 6$ e. $10 \cdot 3 + 6 \cdot 3$ f. $10 \cdot 7 + 9 \cdot 7$
   g. $20 \cdot 6 + 5 \cdot 6$ h. $30 \cdot 8 + 7 \cdot 8$ i. $40 \cdot 5 + 6 \cdot 5$

7. a. Closed for multiplication
   b. Closed for multiplication
   c. Closed for multiplication
   d. Not closed for multiplication; $2 \cdot 3$ is not an element in the set
   e. Closed for multiplication
   f. Closed for multiplication

Appendixes 255

Exercise Set IIe

1. Write each difference as a basic numeral, if the difference exists in the set of whole numbers.

   a. $9 - 2$ b. $11 - 3$ c. $9 - 6$

   d. $7 - 5$ e. $5 - 7$ f. $5 - 12$

   g. $8 - 8$ h. $4 - 12$ i. $8 - 0$

2. Illustrate on a number line the differences in problem 1 which do exist in the set of whole numbers.

3. Write each quotient as a basic numeral, if the quotient exists in the set of whole numbers.

   a. $\dfrac{10}{5}$ b. $\dfrac{21}{7}$ c. $\dfrac{5}{1}$ d. $\dfrac{63}{9}$ e. $\dfrac{12}{5}$

   f. $\dfrac{17}{3}$ g. $\dfrac{1}{7}$ h. $\dfrac{0}{3}$ i. $\dfrac{2}{0}$

4. The set of whole numbers is closed for which of the following operations?

   a. addition   b. subtraction

   c. multiplication   d. division

Answers IIe

1. a. 7   b. 8   c. 3   d. 2   e. Does not exist
   f. Does not exist   g. 0   h. Does not exist   i. 8

2. a. [number line showing 9 forward, 2 back, from 0 to 10]   b. [number line showing 11 forward, 3 back, from 0 to 10]

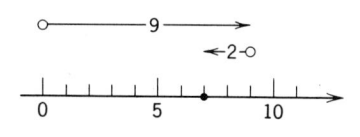

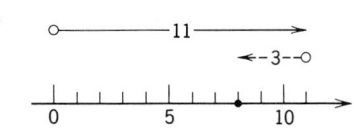

c.

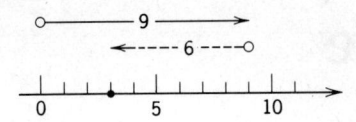

d.

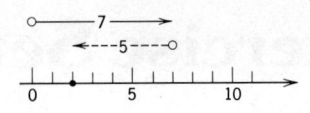

g.

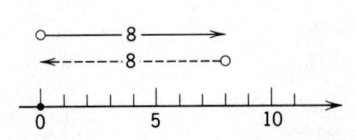

i.

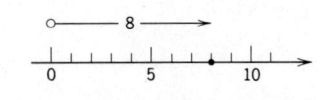

3. a. 2   b. 3   c. 5   d. 7   e. Does not exist
   f. Does not exist   g. Does not exist   h. 0   i. Does not exist

4. Closed for addition and multiplication

Exercise Set IIIa

1. a. If a is a positive integer, then $-a$ is a _____ integer.
   b. If a is a negative integer, then $-a$ is a _____ integer.

2. Graph each set.
   a. $\{2,-4,-6,5\}$
   b. $\{-6,0,2,-4\}$
   c. $\{5,10,15,20\}$
   d. $\{-3,-6,-9,-12\}$
   e. $\{\ldots -2,-1,0,1\}$
   f. $\{-1,0,1,2,\ldots\}$
   g. $\{x|x > 3, x \in J\}$
   h. $\{x|x < -3, x \in J\}$

3. State the additive inverse of each of the following. $a,b \in J$.
   a. 9
   b. 6
   c. -5
   d. -21
   e. 1
   f. a
   g. $-a$
   h. b
   i. $-b$

4. Replace the comma in each pair by $>$ or $<$.
   a. 3, 7
   b. 7, 2
   c. 0, 5
   d. 0, -3
   e. 3, 0
   f. -2, 0
   g. -8, 1
   h. -5, -3
   i. -2, -8

Answers IIIa

1. a. Negative   b. positive

2. a. [number line showing $-6, -5, -4, 0, 2, 5$]
   b. [number line showing $-6, -5, -4, 0, 2$]
   c. [number line showing $0, 5, 10, 15, 20$]

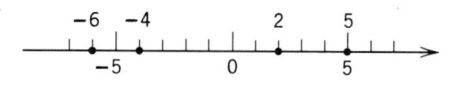

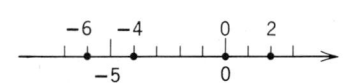

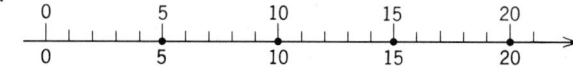

d. 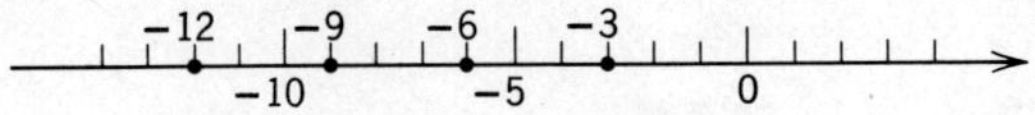

e. 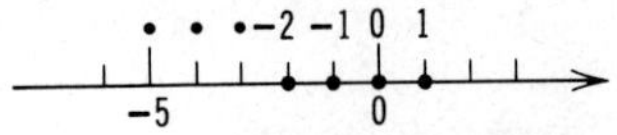

f. 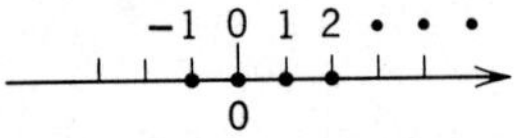

g. 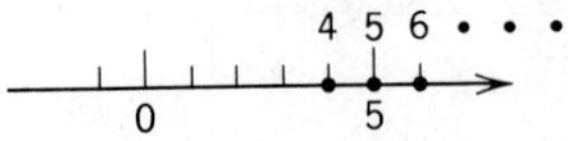

h. 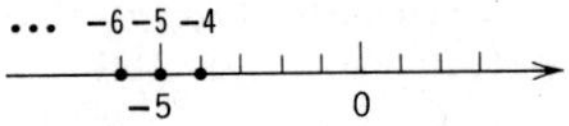

3. a. −9 b. −6 c. 5 d. 21 e. −1
   f. $-a$ g. a h. $-b$ i. b

4. a. $<$ b. $>$ c. $<$ d. $>$ e. $>$
   f. $<$ g. $<$ h. $<$ i. $>$

Exercise Set IIIb

1. a. If a is a positive integer, then $a + a$ is a _____ integer.
   b. If a is a negative integer, then $a + a$ is a _____ integer.

2. Complete the following with the proper symbol, $<$, $=$, or $>$, to form a true statement. $a, b \in J$; $a, b \neq 0$.

   a. $|-4|$ ___ $|4|$ b. $|7|$ ___ $|-2|$ c. $-|-8|$ ___ $|-5|$

   d. $|-2|$ ___ $|3|$ e. $-|4|$ ___ $-|-2|$ f. $-|-6|$ ___ $-|8|$

   g. $-|1|$ ___ $|-6|$ h. $|-a|$ ___ $-|a|$ i. $-|-b|$ ___ $|b|$

3. a. If a is a negative integer and b is a positive integer and $|a| = |b|$, then $a + b = $ _____.

   b. If a is a negative integer and b is a positive integer and $|a| > |b|$, then $a + b$ is a _____ integer.

   c. If a is a negative integer and b is a positive integer and $|a| < |b|$, then $a + b$ is a _____ integer.

4. Write each sum as a basic numeral.

   a. $5 + 2$ b. $-6 + (-2)$ c. $3 + 0$

   d. $0 + (-4)$ e. $8 + (-3)$ f. $-7 + 2$

   g. $-1 + 8$ h. $-7 + 0$ i. $-4 + (-4)$

5. Represent each sum on a number line.

   a. $+2 - 3$ b. $-5 + 3$ c. $0 + 2$

   d. $-6 + (-1)$ e. $-7 + (-4)$ f. $3 + 0$

6. Write each sum as a basic numeral.

   a. $4 + 2 + 3$ b. $-2 + (-3) + (-4)$

   c. $-2 + 0 + (-3)$ d. $5 + 0 + (-5)$

   e. $5 + (-3) + (-7)$ f. $-8 + 2 + (-2)$

   g. $2 + (-9) + (-3)$ h. $-3 + (-1) + 7$

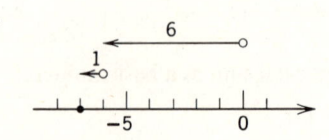

7. Write each difference as a sum and then as a basic numeral.
   a. 7 − 3
   b. 8 − 4
   c. 4 − 6
   d. 8 − 11
   e. 5 − 10
   f. −4 − 2
   g. −12 − 8
   h. −6 − 9
   i. −4 − 8

8. Write each expression as a basic numeral.
   a. −6 − 2 + 2
   b. 4 − 9 + 3
   c. 8 − 1 + 4
   d. 6 − 2 + 7
   e. 8 − 2 + 7
   f. 4 + 0 − 3
   g. −8 − 2 + 7
   h. −2 − 1 − 3
   i. 2 − 9 − 1

Answers IIIb

1. a. Positive   b. negative

2. a. =   b. >   c. <   d. <   e. <
   f. >   g. <   h. >   i. <

3. a. 0   b. negative   c. positive

4. a. 7   b. −8   c. 3   d. −4   e. 5
   f. −5   g. 7   h. −7   i. −8

5. a. [number line]   b. [number line]
   c. [number line]   d. [number line]
   e. [number line]   f. [number line]

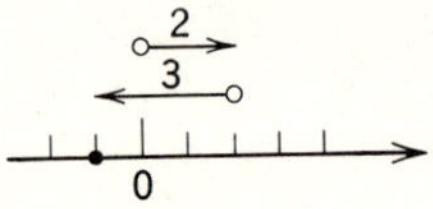

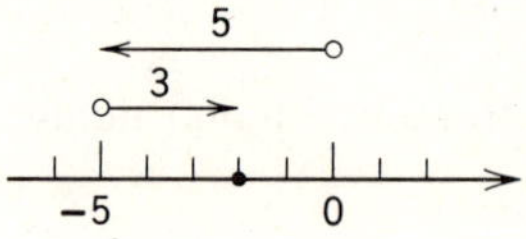

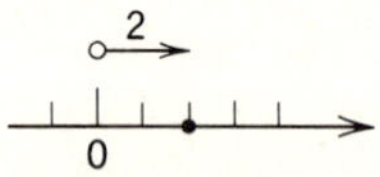

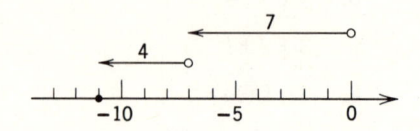

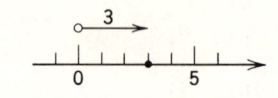

Appendixes

6. a. 9   b. −9   c. −5   d. 0   e. −5
    f. −8   g. −10   h. 3

7. a. $7 + (-3) = 4$ b. $8 + (-4) = 4$ c. $4 + (-6) = -2$
    d. $8 + (-11) = -3$ e. $5 + (-10) = -5$ f. $-4 + (-2) = -6$
    g. $-12 + (-8) = -20$ h. $-6 + (-9) = -15$ i. $-4 + (-8) = -12$

8. a. −6   b. −2   c. 11   d. 11   e. 13
    f. 1   g. −3   h. −6   i. −8

Exercise Set IIIc

1. a. If a and b are both positive integers or both negative integers, then $a \cdot b$ is a _____ integer.

   b. If a is a positive integer and b is a negative integer, then $a \cdot b$ is a _____ integer.

   c. If a is a negative integer and b is a positive integer, then $a \cdot b$ is a _____ integer.

   d. If a is an integer, then $a \cdot 0 =$ _____.

2. Write each product as a basic numeral.

   a. $3 \cdot 4$
   b. $4 \cdot (-5)$
   c. $(-7) \cdot 2$
   d. $(-3) \cdot (-5)$
   e. $(-6) \cdot (-7)$
   f. $(-3) \cdot 0$
   g. $0 \cdot (-5)$
   h. $(-2) \cdot (3) \cdot (-5)$
   i. $(4) \cdot (-3) \cdot (1)$

3. a. If a and b are both positive integers or both negative integers, then $\frac{a}{b}$ is a _____ integer, if such an integer exists.

   b. If a is a positive integer and b is a negative integer, then $\frac{a}{b}$ is a _____ integer, if such an integer exists.

   c. If a is a negative integer and b is a positive integer, then $\frac{a}{b}$ is a _____ integer, if such an integer exists.

   d. If a is zero and b is either a positive or a negative integer, then $\frac{a}{b}$ is _____.

   e. If a is any nonzero integer and $b = 0$, then $\frac{a}{b}$ is _____.

4. Write each quotient as a basic numeral, if the quotient exists in the set of integers.

   a. $\frac{-6}{-3}$
   b. $\frac{-12}{-4}$
   c. $\frac{13}{-6}$
   d. $\frac{-18}{6}$
   e. $\frac{0}{-5}$
   f. $\frac{11}{0}$

▼

Appendixes

g. $\frac{-7}{4}$ h. $\frac{-81}{9}$ i. $\frac{-1}{-1}$

Answers IIIc

1. a. Positive   b. negative   c. negative   d. 0

2. a. 12   b. −20   c. −14   d. 15   e. 42
   f. 0   g. 0   h. 30   i. −12

3. a. Positive   b. negative   c. negative   d. 0
   e. undefined

4. a. 2   b. 3   c. Does not exist   d. −3
   e. 0   f. Does not exist   g. Does not exist
   h. −9   i. 1

Exercise Set IVa

1. Write the multiplicative inverse of each of the following. All variables are elements of J.

   a. 4
   b. 0
   c. -3
   d. -11
   e. $\frac{3}{4}$
   f. $\frac{7}{6}$
   g. $a\,(a \neq 0)$
   h. $a + 5\,(a \neq -5)$
   i. $b - 2\,(b \neq 2)$

2. Write each quotient as a product.

   a. $\frac{2}{3}$
   b. $\frac{3}{7}$
   c. $\frac{4}{5}$
   d. $\frac{12}{5}$
   e. $\frac{7}{3}$
   f. $4 \div 3$
   g. $3 \div 7$
   h. $6 \div 11$
   i. $5 \div 9$

3. Name the axiom represented by each statement. $\frac{a}{b}, \frac{c}{d}, \frac{e}{f} \in Q$.

   a. $\frac{a}{b} + 0 = \frac{a}{b}$
   b. $\frac{a}{b} + \frac{c}{d} = \frac{c}{d} + \frac{a}{b}$
   c. $\frac{a}{b} + \frac{c}{d} \in Q$
   d. If $\frac{a}{b} = \frac{c}{d}$, then $\frac{c}{d} = \frac{a}{b}$
   e. $\frac{a}{b} \cdot \frac{b}{a} = 1$
   f. $\left(\frac{a}{b} \cdot \frac{c}{d}\right) \cdot \frac{e}{f} = \frac{a}{b} \cdot \left(\frac{c}{d} \cdot \frac{e}{f}\right)$
   g. $\frac{a}{b} = \frac{a}{b}$
   h. $\frac{e}{f} \cdot \left(\frac{a}{b} + \frac{c}{d}\right) = \frac{e}{f} \cdot \frac{a}{b} + \frac{e}{f} \cdot \frac{c}{d}$
   i. $\frac{a}{b} \cdot 1 = \frac{a}{b}$
   j. $\frac{a}{b} \cdot \frac{c}{d} = \frac{c}{d} \cdot \frac{a}{b}$
   k. $\frac{a}{b} \cdot \frac{c}{d} \in Q$
   l. $\left(\frac{a}{b} + \frac{c}{d}\right) + \frac{e}{f} = \frac{a}{b} + \left(\frac{c}{d} + \frac{e}{f}\right)$
   m. $\frac{a}{b} + \left(-\frac{a}{b}\right) = 0$
   n. If $\frac{a}{b} = \frac{c}{d}$ and $\frac{c}{d} = \frac{e}{f}$, then $\frac{a}{b} = \frac{e}{f}$.

Appendixes

4. Complete the following (assume no denominator equals 0).

a. If $\frac{a}{b} = \frac{2}{3}$, then $\frac{a}{b} \cdot \frac{1}{2} = \frac{2}{3} \cdot$ _____ .

b. If $\frac{5}{6} = \frac{c}{d}$, then $\frac{5}{6} + \frac{a}{b} = \frac{c}{d} +$ _____ .

c. If $\frac{2}{3} \cdot \frac{c}{d} = \frac{2}{3}$, then $\frac{c}{d} =$ _____ .

d. $\frac{a}{b} \cdot 0 =$ _____ .

e. If $\frac{9}{11} = \frac{c}{d}$, then $\frac{9}{11} \cdot \frac{2}{5} = \frac{c}{d} \cdot$ _____ .

f. If $\frac{3}{4} + \frac{a}{b} = \frac{3}{4}$, then $\frac{a}{b} =$ _____ .

Answers IVa

1. a. $\frac{1}{4}$ b. Does not exist as a rational number

   c. $\frac{1}{-3}$ d. $\frac{1}{-11}$ e. $\frac{4}{3}$ f. $\frac{6}{7}$ g. $\frac{1}{a}$

   h. $\frac{1}{a+5}$ i. $\frac{1}{b-2}$

2. a. $2 \cdot \frac{1}{3}$ b. $3 \cdot \frac{1}{7}$ c. $4 \cdot \frac{1}{5}$ d. $12 \cdot \frac{1}{5}$

   e. $7 \cdot \frac{1}{3}$ f. $4 \cdot \frac{1}{3}$ g. $3 \cdot \frac{1}{7}$ h. $6 \cdot \frac{1}{11}$

   i. $5 \cdot \frac{1}{9}$

3. a. Identity law of addition   b. Commutative law of addition

   c. Closure law for addition   d. Symmetric law of equality

   e. Multiplicative inverse law   f. Associative law of multiplication

▼

g. Reflexive law of equality
i. Identity law of multiplication
k. Closure law for multiplication
m. Additive inverse law

h. Distributive law
j. Commutative law of multiplication
l. Associative law of addition
n. Transitive law of equality

4. a. $\dfrac{2}{3} \cdot \dfrac{1}{2}$ b. $\dfrac{c}{d} + \dfrac{a}{b}$ c. 1

d. 0   e. $\dfrac{c}{d} \cdot \dfrac{2}{5}$ f. 0

Exercise Set IVb

1. Express each of the following as a basic fraction.

a. $\dfrac{6}{8}$ b. $\dfrac{15}{20}$ c. $\dfrac{12}{16}$

d. $\dfrac{9}{15}$ e. $\dfrac{28}{49}$ f. $\dfrac{35}{42}$

g. $\dfrac{22}{33}$ h. $\dfrac{72}{84}$ i. $\dfrac{16}{54}$

2. Graph each set of rational numbers on a number line.

a. $-\dfrac{5}{2}, -\dfrac{3}{2}, \dfrac{3}{2}, \dfrac{7}{2}$ b. $-\dfrac{11}{3}, -\dfrac{7}{3}, -\dfrac{2}{3}, \dfrac{1}{3}$

c. $-\dfrac{11}{4}, -2, 0, \dfrac{5}{4}, \dfrac{9}{4}$ d. $-\dfrac{9}{8}, 0, \dfrac{3}{8}, \dfrac{13}{8}, \dfrac{19}{8}$

3. Replace the comma in each number pair with the proper symbol, = or ≠.

a. $\dfrac{3}{7}, \dfrac{12}{28}$ b. $\dfrac{7}{5}, \dfrac{8}{6}$ c. $\dfrac{0}{5}, \dfrac{2}{7}$

d. $\dfrac{0}{2}, \dfrac{0}{4}$ e. $\dfrac{3}{2}, \dfrac{9}{6}$ f. $\dfrac{4}{11}, \dfrac{20}{55}$

g. $\dfrac{2}{5}, \dfrac{14}{35}$ h. $\dfrac{12}{13}, \dfrac{13}{14}$ i. $\dfrac{13}{15}, \dfrac{39}{45}$

4. Write the missing numerator or denominator so that each second fraction will be equivalent to the first.

a. $\dfrac{3}{4} = \dfrac{?}{8}$ b. $\dfrac{8}{9} = \dfrac{?}{27}$ c. $\dfrac{0}{3} = \dfrac{?}{24}$

d. $\dfrac{0}{-1} = \dfrac{?}{25}$ e. $\dfrac{2}{5} = \dfrac{4}{?}$ f. $\dfrac{1}{6} = \dfrac{3}{?}$

g. $\dfrac{4 \cdot 2}{3 \cdot 5} = \dfrac{4 \cdot 2 \cdot 4}{?}$ h. $\dfrac{1 \cdot 9}{2 \cdot 7} = \dfrac{1 \cdot 9 \cdot 5}{?}$ i. $\dfrac{5 \cdot 0 \cdot 2}{1 \cdot 1 \cdot 1} = \dfrac{?}{1 \cdot 1 \cdot 3}$

5. Which of the following rational numbers are equal to $\frac{20}{30}$?

a. $\frac{20 \cdot 4}{30 \cdot 4}$

b. $\frac{20 - 3}{30 - 3}$

c. $\frac{20 + 7}{30 + 7}$

d. $\frac{20 \cdot (-5)}{30 \cdot (-5)}$

e. $\frac{20 \div 6}{30 \div 6}$

f. $\frac{20 \div 2}{30 \div 3}$

g. $\frac{20 \div 20}{30 \div 30}$

h. $\frac{20 \div \frac{1}{2}}{30 \div \frac{1}{2}}$

i. $\frac{20 \div \frac{2}{3}}{30 \div \frac{2}{3}}$

Answers IVb

1. a. $\frac{3}{4}$ b. $\frac{3}{4}$ c. $\frac{3}{4}$ d. $\frac{3}{5}$ e. $\frac{4}{7}$

   f. $\frac{5}{6}$ g. $\frac{2}{3}$ h. $\frac{6}{7}$ i. $\frac{8}{27}$

2. a.

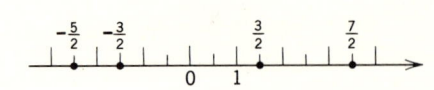

   b.

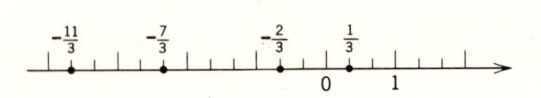

   c.

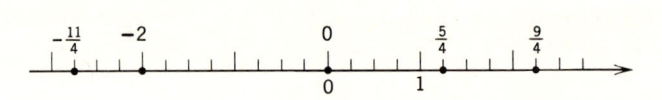

   d.

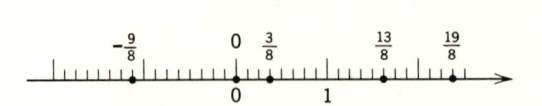

3. a. =   b. ≠   c. ≠   d. =   e. =

   f. =   g. =   h. ≠   i. =

Appendixes

4. a. $\dfrac{3}{4} = \dfrac{6}{8}$ b. $\dfrac{8}{9} = \dfrac{24}{27}$ c. $\dfrac{0}{3} = \dfrac{0}{24}$ d. $\dfrac{0}{-1} = \dfrac{0}{25}$

   e. $\dfrac{2}{5} = \dfrac{4}{10}$ f. $\dfrac{1}{6} = \dfrac{3}{18}$ g. $\dfrac{4 \cdot 2}{3 \cdot 5} = \dfrac{4 \cdot 2 \cdot 4}{3 \cdot 5 \cdot 4} = \dfrac{32}{60}$

   h. $\dfrac{1 \cdot 9}{2 \cdot 7} = \dfrac{1 \cdot 9 \cdot 5}{2 \cdot 7 \cdot 5} = \dfrac{45}{70}$ i. $\dfrac{5 \cdot 0 \cdot 2}{1 \cdot 1 \cdot 1} = \dfrac{5 \cdot 0 \cdot 2 \cdot 3}{1 \cdot 1 \cdot 1 \cdot 3} = \dfrac{0}{3} = 0$

5. a. Equal   b. not equal   c. not equal
   d. equal   e. equal   f. not equal
   g. not equal   h. equal   i. equal

Exercise Set IVc

1. Express each sum as a basic fraction.

a. $\dfrac{4}{7} + \dfrac{2}{7}$ b. $\dfrac{2}{9} + \dfrac{3}{9}$ c. $\dfrac{3}{5} + \dfrac{1}{5}$

d. $\dfrac{7}{11} + \dfrac{3}{11}$ e. $\dfrac{2}{3} + \dfrac{5}{3} + \dfrac{1}{3}$ f. $\dfrac{4}{7} + \dfrac{2}{7} + \dfrac{1}{7}$

g. $\dfrac{8}{9} + \dfrac{2}{9} + \dfrac{1}{9}$ h. $\dfrac{3}{17} + \dfrac{1}{17} + \dfrac{5}{17}$ i. $\dfrac{1}{16} + \dfrac{3}{16} + \dfrac{5}{16}$

2. Express each sum as a basic fraction.

a. $\dfrac{2}{5} + \dfrac{3}{10}$ b. $\dfrac{1}{3} + \dfrac{5}{6}$ c. $\dfrac{2}{3} + \dfrac{3}{5}$

d. $\dfrac{3}{5} + \dfrac{2}{7}$ e. $\dfrac{1}{4} + \dfrac{3}{11}$ f. $\dfrac{1}{5} + \dfrac{2}{9}$

g. $\dfrac{1}{5} + \dfrac{3}{10} + \dfrac{2}{15}$ h. $\dfrac{2}{3} + \dfrac{1}{6} + \dfrac{5}{18}$ i. $\dfrac{1}{6} + \dfrac{3}{5} + \dfrac{5}{4}$

3. Write each expression as a basic fraction.

a. $2\dfrac{1}{3}$ b. $3\dfrac{1}{7}$ c. $4\dfrac{2}{5}$

d. $3\dfrac{4}{7}$ e. $6\dfrac{5}{6}$ f. $2\dfrac{6}{7}$

g. $1\dfrac{3}{11}$ h. $4\dfrac{2}{15}$ i. $5\dfrac{8}{9}$

4. Express each sum as a basic fraction.

a. $1\dfrac{1}{3} + 1\dfrac{3}{4}$ b. $1\dfrac{2}{3} + 2\dfrac{1}{5}$ c. $2\dfrac{1}{5} + 3\dfrac{1}{3}$

d. $3\dfrac{2}{3} + 4\dfrac{1}{5}$ e. $3 + 5\dfrac{1}{4}$ f. $2\dfrac{5}{6} + 5$

g. $\dfrac{1}{2} + 2\dfrac{1}{3}$ h. $\dfrac{1}{3} + 7\dfrac{1}{5}$ i. $4\dfrac{3}{8} + \dfrac{2}{3}$

Appendixes

Answers IVc

1. a. $\dfrac{6}{7}$ b. $\dfrac{5}{9}$ c. $\dfrac{4}{5}$ d. $\dfrac{10}{11}$ e. $\dfrac{8}{3}$

   f. $\dfrac{7}{7}=1$ g. $\dfrac{11}{9}$ h. $\dfrac{9}{17}$ i. $\dfrac{9}{16}$

2. a. $\dfrac{7}{10}$ b. $\dfrac{7}{6}$ c. $\dfrac{19}{15}$ d. $\dfrac{31}{35}$ e. $\dfrac{23}{44}$

   f. $\dfrac{19}{45}$ g. $\dfrac{19}{30}$ h. $\dfrac{20}{18}=\dfrac{10}{9}$ i. $\dfrac{121}{60}$

3. a. $\dfrac{7}{3}$ b. $\dfrac{22}{7}$ c. $\dfrac{22}{5}$ d. $\dfrac{25}{7}$ e. $\dfrac{41}{6}$

   f. $\dfrac{20}{7}$ g. $\dfrac{14}{11}$ h. $\dfrac{62}{15}$ i. $\dfrac{53}{9}$

4. a. $\dfrac{37}{12}$ b. $\dfrac{58}{15}$ c. $\dfrac{83}{15}$ d. $\dfrac{118}{15}$ e. $\dfrac{33}{4}$

   f. $\dfrac{47}{6}$ g. $\dfrac{17}{6}$ h. $\dfrac{113}{15}$ i. $\dfrac{121}{24}$

Exercise Set IVd

1. Express each product as a basic fraction.

   a. $\dfrac{2}{3} \cdot \dfrac{4}{7}$
   b. $\dfrac{1}{5} \cdot \dfrac{3}{2}$
   c. $\dfrac{4}{5} \cdot \dfrac{1}{3}$

   d. $\dfrac{3}{7} \cdot \dfrac{2}{5}$
   e. $\dfrac{3}{5} \cdot \dfrac{2}{3}$
   f. $\dfrac{7}{9} \cdot \dfrac{3}{4}$

   g. $\dfrac{4}{9} \cdot 3$
   h. $6 \cdot \dfrac{4}{7}$
   i. $\dfrac{5}{2} \cdot 16$

2. Write each product as a basic fraction.

   a. $2\dfrac{1}{3} \cdot \dfrac{1}{2}$
   b. $\dfrac{1}{4} \cdot 3\dfrac{1}{2}$
   c. $\dfrac{2}{3} \cdot 4\dfrac{2}{5}$

   d. $2\dfrac{3}{7} \cdot \dfrac{2}{3}$
   e. $1\dfrac{1}{2} \cdot 2\dfrac{1}{2}$
   f. $3\dfrac{1}{4} \cdot 5\dfrac{1}{2}$

   g. $6\dfrac{1}{2} \cdot 3\dfrac{2}{5}$
   h. $2 \cdot 4\dfrac{1}{3}$
   i. $6\dfrac{3}{4} \cdot 3$

3. Write each product as a basic fraction.

   a. $\dfrac{2}{5} \cdot \dfrac{1}{7} \cdot \dfrac{3}{5}$
   b. $\dfrac{1}{6} \cdot \dfrac{5}{3} \cdot \dfrac{7}{2}$
   c. $\dfrac{1}{2} \cdot \dfrac{1}{2} \cdot \dfrac{1}{2}$

   d. $\dfrac{1}{3} \cdot \dfrac{1}{3} \cdot \dfrac{1}{3}$
   e. $\dfrac{2}{3} \cdot \dfrac{3}{4} \cdot \dfrac{4}{5}$
   f. $\dfrac{3}{4} \cdot \dfrac{4}{5} \cdot \dfrac{5}{6}$

   g. $\dfrac{2}{3} \cdot \dfrac{4}{7} \cdot \dfrac{1}{9}$
   h. $\dfrac{3}{4} \cdot \dfrac{1}{11} \cdot \dfrac{5}{7}$
   i. $\dfrac{4}{5} \cdot \dfrac{2}{9} \cdot \dfrac{3}{13}$

4. Write each expression as a basic fraction.

   a. $\dfrac{1}{3}\left(\dfrac{2}{5} + \dfrac{1}{4}\right)$
   b. $\dfrac{1}{2}\left(\dfrac{3}{4} + \dfrac{3}{5}\right)$
   c. $\dfrac{1}{4}\left(\dfrac{1}{6} + \dfrac{2}{7}\right)$

   d. $\dfrac{1}{5}\left(\dfrac{1}{3} + \dfrac{3}{7}\right)$
   e. $\dfrac{1}{3}\left(\dfrac{2}{7} + \dfrac{1}{5}\right)$
   f. $\dfrac{1}{2}\left(\dfrac{5}{8} + \dfrac{1}{3}\right)$

   g. $\dfrac{1}{6}\left(\dfrac{3}{5} + \dfrac{7}{9}\right)$
   h. $\dfrac{1}{7}\left(\dfrac{2}{5} + \dfrac{3}{11}\right)$
   i. $\dfrac{1}{8}\left(\dfrac{3}{7} + \dfrac{5}{12}\right)$

Appendixes

Answers IVd

1. a. $\dfrac{8}{21}$ b. $\dfrac{3}{10}$ c. $\dfrac{4}{15}$ d. $\dfrac{6}{35}$ e. $\dfrac{6}{15} = \dfrac{2}{5}$
   f. $\dfrac{21}{36} = \dfrac{7}{12}$ g. $\dfrac{12}{9} = \dfrac{4}{3}$ h. $\dfrac{24}{7}$ i. $\dfrac{80}{2} = 40$

2. a. $\dfrac{7}{6}$ b. $\dfrac{7}{8}$ c. $\dfrac{44}{15}$ d. $\dfrac{34}{21}$ e. $\dfrac{15}{4}$
   f. $\dfrac{143}{8}$ g. $\dfrac{221}{10}$ h. $\dfrac{26}{3}$ i. $\dfrac{81}{4}$

3. a. $\dfrac{6}{175}$ b. $\dfrac{35}{36}$ c. $\dfrac{1}{8}$ d. $\dfrac{1}{27}$ e. $\dfrac{24}{60} = \dfrac{2}{5}$
   f. $\dfrac{60}{120} = \dfrac{1}{2}$ g. $\dfrac{8}{189}$ h. $\dfrac{15}{308}$ i. $\dfrac{24}{585} = \dfrac{8}{195}$

4. a. $\dfrac{13}{60}$ b. $\dfrac{27}{40}$ c. $\dfrac{19}{168}$ d. $\dfrac{16}{105}$ e. $\dfrac{17}{105}$
   f. $\dfrac{23}{48}$ g. $\dfrac{62}{270} = \dfrac{31}{135}$ h. $\dfrac{37}{385}$ i. $\dfrac{71}{672}$

Exercise Set IVe

1. Express each difference as a basic fraction.

   a. $\dfrac{4}{3} - \dfrac{2}{3}$ b. $\dfrac{7}{4} - \dfrac{5}{4}$ c. $\dfrac{5}{7} - \dfrac{2}{7}$

   d. $\dfrac{7}{9} - \dfrac{5}{9}$ e. $\dfrac{7}{10} - \dfrac{1}{2}$ f. $\dfrac{4}{9} - \dfrac{1}{3}$

   g. $\dfrac{5}{6} - \dfrac{2}{3}$ h. $\dfrac{7}{12} - \dfrac{3}{8}$ i. $\dfrac{7}{10} - \dfrac{1}{15}$

2. Express each difference as a basic fraction.

   a. $3\dfrac{3}{2} - 2\dfrac{2}{3}$ b. $5\dfrac{3}{4} - 1\dfrac{1}{3}$ c. $4\dfrac{3}{5} - \dfrac{2}{7}$

   d. $2\dfrac{4}{9} - \dfrac{1}{8}$ e. $4 - \dfrac{1}{7}$ f. $5 - \dfrac{3}{4}$

   g. $1\dfrac{3}{4} - \dfrac{7}{8}$ h. $3\dfrac{1}{5} - 2\dfrac{3}{4}$ i. $2\dfrac{2}{3} - 1\dfrac{1}{6} - \dfrac{2}{5}$

3. Write each quotient as a basic fraction.

   a. $\dfrac{1}{4} \div \dfrac{1}{2}$ b. $\dfrac{3}{2} \div \dfrac{1}{4}$ c. $\dfrac{3}{5} \div \dfrac{1}{10}$

   d. $\dfrac{3}{8} \div \dfrac{1}{12}$ e. $\dfrac{3}{5} \div \dfrac{1}{2}$ f. $\dfrac{1}{7} \div \dfrac{2}{3}$

   g. $\dfrac{5}{9} \div \dfrac{2}{7}$ h. $\dfrac{11}{3} \div \dfrac{4}{5}$ i. $\dfrac{10}{13} \div \dfrac{1}{7}$

4. Write each quotient as a basic fraction.

   a. $4 \div \dfrac{2}{3}$ b. $5 \div \dfrac{7}{8}$ c. $\dfrac{4}{9} \div 2$

   d. $\dfrac{3}{5} \div 6$ e. $2\dfrac{1}{2} \div \dfrac{3}{4}$ f. $\dfrac{2}{5} \div 4\dfrac{1}{3}$

   g. $6\dfrac{1}{2} \div 2\dfrac{1}{4}$ h. $4 \div 2\dfrac{3}{5}$ i. $2\dfrac{1}{5} \div 6$

Appendixes

5. Write each expression as a basic fraction.

a. $\left(\dfrac{1}{4} + \dfrac{1}{3}\right) \div \dfrac{1}{5}$
b. $\left(\dfrac{2}{5} + \dfrac{1}{3}\right) \div \dfrac{1}{4}$
c. $\left(\dfrac{3}{4} + \dfrac{2}{3}\right) \div \dfrac{2}{5}$

d. $\left(\dfrac{3}{5} + \dfrac{3}{4}\right) \div \dfrac{3}{7}$
e. $\dfrac{5}{7} \div \left(\dfrac{2}{3} - \dfrac{3}{7}\right)$
f. $\dfrac{4}{9} \div \left(\dfrac{5}{7} - \dfrac{2}{5}\right)$

g. $\left(\dfrac{1}{2} - \dfrac{1}{3}\right) \div \left(\dfrac{1}{2} + \dfrac{1}{3}\right)$
h. $\left(\dfrac{4}{5} - \dfrac{1}{3}\right) \div \left(\dfrac{1}{2} + \dfrac{1}{5}\right)$
i. $\left(\dfrac{4}{9} - \dfrac{1}{4}\right) \div \left(\dfrac{3}{2} - \dfrac{5}{7}\right)$

Answers IVe

1. a. $\dfrac{2}{3}$
b. $\dfrac{2}{4} = \dfrac{1}{2}$
c. $\dfrac{3}{7}$
d. $\dfrac{2}{9}$
e. $\dfrac{2}{10} = \dfrac{1}{5}$
f. $\dfrac{1}{9}$
g. $\dfrac{1}{6}$
h. $\dfrac{5}{24}$
i. $\dfrac{19}{30}$

2. a. $\dfrac{11}{6}$
b. $\dfrac{53}{12}$
c. $\dfrac{151}{35}$
d. $\dfrac{167}{72}$
e. $\dfrac{27}{7}$
f. $\dfrac{17}{4}$
g. $\dfrac{7}{8}$
h. $\dfrac{9}{20}$
i. $\dfrac{11}{10}$

3. a. $\dfrac{1}{2}$
b. $\dfrac{12}{2} = 6$
c. $\dfrac{30}{5} = 6$
d. $\dfrac{36}{8} = \dfrac{9}{2}$
e. $\dfrac{6}{5}$
f. $\dfrac{3}{14}$
g. $\dfrac{35}{18}$
h. $\dfrac{55}{12}$
i. $\dfrac{70}{13}$

4. a. $\dfrac{12}{2} = 6$
b. $\dfrac{40}{7}$
c. $\dfrac{4}{18} = \dfrac{2}{9}$
d. $\dfrac{3}{30} = \dfrac{1}{10}$
e. $\dfrac{20}{6} = \dfrac{10}{3}$
f. $\dfrac{6}{65}$
g. $\dfrac{52}{18} = \dfrac{26}{9}$
h. $\dfrac{20}{13}$
i. $\dfrac{11}{30}$

5. a. $\dfrac{35}{12}$
b. $\dfrac{44}{15}$
c. $\dfrac{85}{24}$
d. $\dfrac{189}{60} = \dfrac{63}{20}$
e. $\dfrac{105}{35} = 3$
f. $\dfrac{140}{99}$
g. $\dfrac{6}{30} = \dfrac{1}{5}$
h. $\dfrac{70}{105} = \dfrac{2}{3}$
i. $\dfrac{98}{396} = \dfrac{49}{198}$

Exercise Set Va

1. Write each square root as a basic numeral or as a basic fraction.

   a. $\sqrt{4}$
   b. $\sqrt{36}$
   c. $\sqrt{49}$
   d. $\sqrt{81}$
   e. $-\sqrt{1}$
   f. $-\sqrt{64}$
   g. $\sqrt{\dfrac{9}{25}}$
   h. $\sqrt{\dfrac{16}{121}}$
   i. $-\sqrt{\dfrac{100}{9}}$

2. Graph each set of numbers. Use decimal approximations on page 241 to approximate the position of the graphs of irrational numbers.

   a. $\{-\sqrt{11}, 0, \sqrt{3}\}$
   b. $\{-\sqrt{21}, -2, \sqrt{4}\}$
   c. $\{-5, -\sqrt{2}, \sqrt{29}\}$
   d. $\{-\sqrt{70}, -\sqrt{17}, \sqrt{51}\}$
   e. $\{-\sqrt{6}, \sqrt{7}, \sqrt{40}\}$
   f. $\{-\sqrt{44}, -\sqrt{15}, 3\}$

3. Graph each set of numbers.

   a. $\{a | -2 < a < 1, a \in R\}$
   b. $\{a | 3 < a < 7, a \in R\}$
   c. $\{a | -3 \leqslant a < 3, a \in R\}$
   d. $\{a | -2 < a \leqslant 4, a \in R\}$
   e. $\{a | 0 \leqslant a \leqslant 5, a \in R\}$
   f. $\{a | -6 \leqslant a \leqslant 0, a \in R\}$

4. Which of the following statements are true for every value of the variable in its replacement set?

   a. If $a \in N$, then $a \in W$.
   b. If $a \in W$, then $a \in N$.
   c. If $a \in Q$, then $a \in N$.
   d. If $a \in W$, then $a \in Q$.
   e. If $a \in W$, then $a \in H$.
   f. If $a \in J$, then $a \in W$.
   g. If $a \in Q$, then $a \in W$.
   h. If $a \in J$, then $a \in N$.

Answers Va

1. a. 2   b. 6   c. 7   d. 9   e. -1
   f. -8 g. $\dfrac{3}{5}$ h. $\dfrac{4}{11}$ i. $-\dfrac{10}{3}$

Appendixes

2. a. [number line showing $-\sqrt{11}$, 0, $\sqrt{3}$, 5]

   b. [number line showing $\sqrt{21}$, -2, $\sqrt{4}$, -5, 0]

   c. [number line showing -5, $-\sqrt{2}$, $\sqrt{29}$, -5, 0, 5]

   d. [number line showing $-\sqrt{70}$, $-\sqrt{17}$, $\sqrt{51}$, -5, 0, 5]

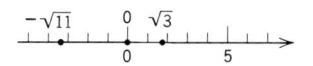

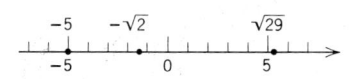

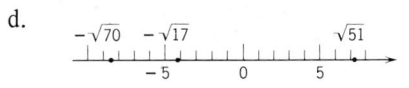

   e. [number line showing $-\sqrt{6}$, $\sqrt{7}$, $\sqrt{40}$, 0, 5]

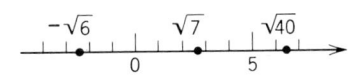

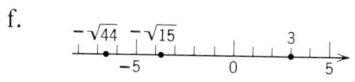

   f. [number line showing $-\sqrt{44}$, $-\sqrt{15}$, 3, -5, 0, 5]

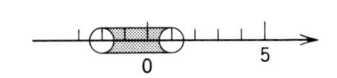

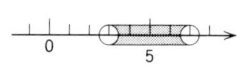

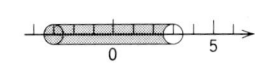

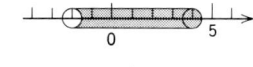

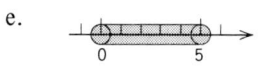

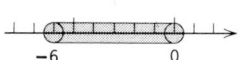

3. a. [shaded interval on number line, 0 to 5]

   b. [shaded interval on number line, 0 to 5]

   c. [shaded interval on number line, 0 to 5]

   d. [shaded interval on number line, 0 to 5]

   e. [shaded interval on number line, 0 to 5]

   f. [shaded interval on number line, -6 to 0]

4. a. true   b. false   c. false   d. true

   e. false   f. false   g. false   h. false

Answers to Self-Evaluation Tests

UNIT I

1. element (or member)
2. subset
3. intersection
4. union
5. ∅
6. ⊂
7. U
8. ≠
9. are
10. 4
11. one-to-one
12. equivalent to
13. are not
14. {a, b, c, d}
15. {c}
16. {c}
17. A ∪ B
18. A ∩ B
19. {(c, g), (c, e), (f, g), (f, e)}
20. {(○, △), (□, △)}

UNIT II

1. infinite
2. less than
3. ordinal
4. variable
5. coordinate
6. Reflexive
7. Symmetric
8. Transitive
9. axiom
10. Transitive
11. sum
12. closed
13. Associative
14. Commutative
15. Identity
16. theorems
17. always
18. A × B
19. 15
20. lattice
21. identity
22. Closure
23. Associative
24. always
25. always
26. always
27. never
28. $b + d = a$
29. sometimes
30. $b \cdot q = a$

UNIT III

1. integers
2. W ⊂ J
3. infinite
4. additive inverses
5. sometimes
6. $-9 < -5 < 7$
7. always
8. c
9. less than
10. always
11. always
12. always
13. absolute value

278

Appendixes

14. sometimes **15.** a **16.** 2 **17.** never **18.** always
19. 56 **20.** zero **21.** always **22.** always **23.** sometimes
24. 0 **25.** undefined

UNIT IV

1. multiplicative inverse **2.** is **3.** does **4.** always
5. [number line with $\frac{5}{3}$ marked between 0 and 1 — past 1] **6.** always **7.** does not

8. $\frac{4}{5}$ **9.** $\frac{31}{35}$ **10.** $\frac{2}{3}$ **11.** $\frac{1}{15}$ **12.** always **13.** $\frac{1}{6}$
14. [array of dots with 2×3 rectangle outlined in upper left of a 5×7 grid] **15.** $\frac{4}{15}$ **16.** always **17.** $\frac{a}{b} \cdot \frac{d}{c}$
18. $\frac{28}{15}$ **19.** $\frac{10}{13}$ **20.** always

UNIT V

1. square root **2.** b **3.** 6 **4.** 7 **5.** irrational
6. [number line showing $-\sqrt{36}$, $-\sqrt{5}$, $\sqrt{34}$ marked between −5, 0, 5] **7.** $R; \emptyset$ **8.** completeness

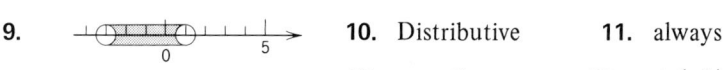

9. [number line with shaded segment from 0 to about 5] **10.** Distributive **11.** always
12. sometimes **13.** $a+(-b)$ **14.** always
15. always **16.** $a \cdot \frac{1}{b}$ **17.** $\frac{a}{b} \cdot \frac{d}{c}$
18. N,W,H **19.** N,W **20.** N,W,J

Index

(The page on which the item first appears is shown.)

Absolute value, 120
Addition algorithm, 60
Additive inverse law, 110
Array, 65
Associative law
    of addition, 53
    of multiplication, 69
Axiom, 47

Basic fraction, 170
Basic numeral, 58
Binary operation, 54

Cardinal use, 32
Cartesian product, 23
Closure law
    for addition, 52
    for multiplication, 68
Commutative law
    of addition, 53
    of multiplication, 68
Complement of a set, 15
Completeness property, 217
Coordinates of a point, 42
Counting numbers, 37

Denominator of a fraction, 152
Differences
    of integers, 127
    of rational numbers, 189
    of whole numbers, 88

Element of a set, 3
Empty set, 6
Equal sets, 4
Equivalent sets, 10

Finite sets, 37
Fraction
    basic, 170
    denominator of, 152
    fundamental principle of, 165
    improper, 178
    numerator of, 152
    proper, 178
    reducing of a, 168
    terms of, 167
Fundamental principle of fractions, 165

Graphs
    of integers, 108
    of numbers, 41
    of rational numbers, 158
    of real numbers, 220
    of whole numbers, 40
Greater than, 35

Identity element, 55
Identity law
    of addition, 55
    of multiplication, 67
Improper fraction, 178
Infinite set, 37
Integers, 106
    differences of, 127
    graph of, 108
    product of, 130
    quotient of, 137
    set of, 109
    sum of, 117
Intersection of sets, 13
Irrational numbers, 215

281

Lattice, 65
Least common denominator, 175
Less than, 34

Member of a set, 3
Multiplicative inverse law, 147

Natural number(s), 34
    product of, 62
    sum of, 51
Negative integers, 107
Negative law, 110
Negative numbers, 107
Negative of a number, 107
Null set, 6
Number(s)
    graph of, 41
    irrational, 215
    natural, 34
    negative, 107
    negative of a, 107
    positive, 108
    rational, 146
    real, 212
    whole, 30
Number line, 40
Number system, 84
Numeral, 33
Numerator of a fraction, 152

One-to-one correspondence, 9
Ordered pairs, 23
Ordered set, 35
Ordinal use, 36

Positive number, 108
Product
    of integers, 130
    of natural numbers, 62
    of rational numbers, 188
    of whole numbers, 51
Proof, 56
Proper fractions, 178

Quotient
    of integers, 137
    of rational numbers, 192
    of whole numbers, 88

Radical expressions, 214
Radicand, 214

Rational numbers
    differences of, 189
    graph of, 158
    product of, 188
    quotient of, 192
    sum of, 179
Real numbers
    graph of, 220
    set of, 216
Reciprocal, 148
Reducing a fraction, 168
Reflexive law of equality, 46
Replacement set, 38

Scale numbers, 40
Set(s)
    complement of, 15
    disjoint, 11
    element of, 3
    equal, 4
    equivalent, 10
    finite, 37
    infinite, 37
    intersection of, 13
    member of, 3
    of integers, 109
    of real numbers, 216
    of whole numbers, 34
    ordered, 35
    union of, 12
Set-builder notation, 39
Signed numbers, 112
Square root, 213
Subsets, 7
Substitution law of equality, 48
Sum
    of integers, 117
    of natural numbers, 51
    of rational numbers, 179
    of whole numbers, 51
Symmetric law of equality, 46

Terms of the fraction, 167
Theorems, 56
Transitive law
    of equality, 46
    of inequality, 49
Trichotomy law, 48

Union of sets, 12
Universe, 14

Index

Variable, 38
Venn diagrams, 16

Whole numbers, 30
    differences of, 85
    graph of, 40

Whole numbers (*continued*)
    product of, 51
    quotient of, 88
    set of, 34
    sum of, 51